COUNT DOWN
TO GLORY
NASA'S TRIALS AND TRIUMPHS IN SPACE

KENT ALEXANDER

PRICE STERN SLOAN
Los Angeles

A FRIEDMAN GROUP BOOK

© 1989 by Michael Friedman Publishing Group, Inc.

Published by Price Stern Sloan, Inc.
360 North La Cienega Boulevard, Los Angeles, California 90048

Printed in Hong Kong.

9 8 7 6 5 4 3 2 1

First Printing.

LIBRARY OF CONGRESS CATALOGING-IN-PUBLICATION DATA

Alexander, Kent
    Countdown to glory / by Kent Alexander.
        p.   cm.
    ISBN 0-89586-787-7
    1. United States.   National Aeronautics and Space Administration—
-History.   I. Title.
TL521.312.A626   1989   88-31657
353.0087'78—dc19   CIP

COUNTDOWN TO GLORY: NASA's Trials and Triumphs in Space
was prepared and produced by
Michael Friedman Publishing Group, Inc.
15 West 26th Street
New York, New York 10010

Editor: Sharon Kalman
Art Director: Robert W. Kosturko
Designer: Alanna Georgens
Photography Editor: Christopher Bain
Photo Researcher: Daniella Jo Nilva
Production Manager: Karen L. Greenberg

All photographs courtesy of NASA

Typeset by Mar + x Myles Graphics, Inc.
Color separations by South Sea International Press, Ltd.
Printed and bound in Hong Kong by Leefung-Asco Printers, Ltd.

DEDICATION

This book is dedicated to Terry Buerkle whose patience permitted me to smile through this massive project; to those men and women who have dreamed the impossible and had the strength to pursue it; and to our future, which will, hopefully, be even more fruitful than our past.

ACKNOWLEDGMENTS

A hearty thanks to the people of the New York Public Library, the wonderful employees of NASA, and to my editor, Sharon Kalman.

Table of Contents

Onlookers watch as Orville and Wilbur Wright test the first airplane, at Kitty Hawk, North Carolina.

CHAPTER 1

"The Earth is the cradle of humanity, but mankind will not stay in the cradle forever."

—Konstantin Tsiolkovsky

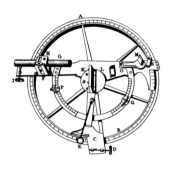

GERMINATION

The history of NASA encompasses more than just the Apollo program, the Space Shuttle and the Space Station. It also includes the early history of the universe, of space flight—rich and exciting in itself—and of rocketry, which may date further back than the thirteenth century.

The first steps toward a scientific view of the universe were taken in Babylonia and Egypt at least 5,000 years ago. As early as 3000 B.C., Babylonian astrologer/astronomers were making methodical observations of the heavens.

By the second millennium, astronomers had fit the planets to the system of the zodiac, and by 1000 B.C., they were apparently keeping records of the movements of the brighter planets, as well as the Sun and the Moon. Tables of the motion of Venus between 1921 and 1901 B.C. have been found by archaeologists, and data on Mars and Jupiter were catalogued for future reference.

As astonishing as these achievements were, the Greek philosopher Heraclides of Pontus (c. 388–315 B.C.) theorized that the daily rotation of the stars was due to the Earth turning on its axis. He also discovered that Mercury and Venus revolved around the Sun, not the Earth. These observations from this forward-thinking man opened the gates to one of the great intellectual jumps in human history.

Aristarchus of Samos (c. 310–230 B.C.) went a crucial step beyond Heraclides. He said the Earth revolved around the Sun! This revolutionary idea was ridiculed and then discarded; it was not accepted until nearly 2,000 years later. Only Seleucus of Seleucia, a Chaldean who lived 100 years later than Aristarchus, accepted this theory as true.

It has taken nearly 5,000 years to accumulate what we currently know about our universe. A trip to the Moon could not be realistically imagined until astronomers were able to give an accurate picture of the universe. One of the earliest fictional descriptions of space travel was written by Lucian of Samosata, a Greek sophist and satirist, in the second century A.D. His *Vera Historia* (True History) is probably the first space fiction written; it includes a trip through space, a landing on another planet, a description of that world and a return to Earth. This voyage was accomplished in a sailing vessel that was lifted from the sea by a violent whirlwind. As man's astronomical knowledge increased, the fictional space voyages devised by restless imaginations became correspondingly more sophisticated.

The science of rocketry dates to the thirteenth century, when black-powder (charcoal, sulphur and saltpeter) rockets were used. The Chinese are generally credited with being the first to use a propulsive device, although there are no hard facts to support this. It is accepted by historians, however, that crude rockets were fired as military weapons in the thirteenth century in the Far East, the Near East and Europe.

The effect of "reaction propulsion" was known much earlier to the Greeks,

Above, top: An early drawing of the world as it was conceived by the Chaldeans. *Above, below:*

A drawing of the ancient Greek mathematician and inventor, Archimedes, who lived between

287 and 212 B.C.

Tycho Brahe's Quadrant (*below, left*) for studying the placement of the stars. Brahe, an early Danish astronomer, lived from 1546 to 1601. *Left:* Sir Isaac Newton, English physicist, astronomer and mathematician. *Below, bottom:* Jules Verne, the science fiction writer who inspired generations of scientists, astronomers and inventors. *Below:* A detailed drawing of Henson's flying machine, which predates 1850.

Opposite page: The Army-JPL (Jet Propulsion Lab) Bumper Wac begins its ascent from its White Sands testing ground launch site in New Mexico. Left: This detailed drawing of early weapons includes cannons and rockets. The drawing dates to approximately 1850.

but the principles involved in this reaction were apparently not studied or developed any further.

So it is that as astronauts set foot on the Moon and on distant planets, they owe a debt to writers like Jules Verne, who helped fire our scientific imagination, as well as to scientists like Sir Isaac Newton (see page 11), who prepared the way for actually getting there. While writers of science fiction used boats, birds, bottles, chariots and spirits to carry their heroes to the Moon and beyond, the device that would finally make the trip possible was undergoing a slow but steady evolution—from a curiosity to a toy to a weapon, from a crude device to a sophisticated machine. Eventually, the time was to come when the rocket would take its place as our only means of space transportation.

THE ROCKET'S RED GLARE

The rocket is a reaction device that works in accordance with Sir Isaac Newton's third law of motion: For every action there is an equal and opposite reaction.

In the beginning of the nineteenth century, Sir William Concreve (1772–1828) developed war rockets that could be fired from land or sea. These rockets were used by the British Army in such diverse places as Copenhagen, Denmark, Bologna, Italy and, later, the Potomac River.

Around 1839, William Hale, a self-educated naval designer and ordnance expert living in Greenwich, Connecticut, directed his energies and abilities toward creating a rocket. Within a few years he had devised a scheme whereby he could give rotary motion to a rocket by directing part of the exhaust flame through slanted, peripheral openings. In 1844, he was awarded a patent for this "stickless" or "rotary" rocket, and in 1863 he published *A Treatise on the Comparative Merits of a Rifle Gun and Rotary Rocket.* In 1846, Hale sold his invention, a spin-stabilized rocket, to the United States Army Ordnance Department. Throughout his lifetime, Hale improved upon launchers for both land and naval rockets.

By the middle of the nineteenth century, every major nation in Europe had rocket brigades. Rockets were launched several times by the United States in the Mexican War in 1846–47, and by both the Union and Confederate armies during the Civil War.

Right: December 17, 1903, 10:30 a.m. This airplane, invented by Orville and Wilbur Wright of Dayton, Ohio, took off from Kitty Hawk, North Carolina, and flew straight into aviation history. *Below:* A NACA employee in a Ford Model T truck with a "huck starter" prepares to start a Lewis & Vought VE-7 research plane. *Below right:* Orville and Wilbur Wright discuss the intricacies of their Wright Flyer.

By 1900 Jules Verne, the science fiction writer who inspired a generation of scientists, had already seen many of his fictional ideas become technical fact at the Paris World's Fair in 1889. Meanwhile, Orville and Wilbur Wright, two brothers who owned a Dayton, Ohio, bicycle shop, were preparing to test their newly invented glider near Kitty Hawk, North Carolina.

On December 17, 1903, at 10:35 A.M., in a twenty-seven mph (forty-four kph) wind, Orville Wright made aviation history in the first engine-powered flight. This flight lasted twelve seconds and traveled a distance of 120 feet (36.6 meters). The news of the flying machine with a homemade twelve horsepower engine quietly spread; soon their feat was well-known, and they eventually sold their Military Flyer to the United States Army.

Although Europe didn't see its first powered flight until 1906, at the dawn of World War I every major nation had aeronautical research facilities except the United States.

NACA—NASA'S PREDECESSOR

The lack of a government laboratory devoted to the science of flight prompted the creation of the National Committee on Aeronautics (NACA), NASA's predecessor. Founded in 1915 just before the United States entered World War I, NACA's goal was to bring the backward state of American aviation to the level of the European countries.

At its inception NACA was a service agency. Its primary responsibility was scientific lab research in aeronautics; it both served the needs of all government departments and coordinated aeronautic research. Through membership on committees and subcommittees, NACA was not only linked to government agencies concerned with flight, but also to the aviation and allied industries and—of primary concern—to educational and scientific institutions. Through sponsored research, symposiums, technical conferences and reports, it also distributed research and information.

On September 3, 1908, the Wright Flyer, designed by Orville and Wilbur Wright, is demonstrated at Fort Meyer, Virginia. This was the first United States military aircraft.

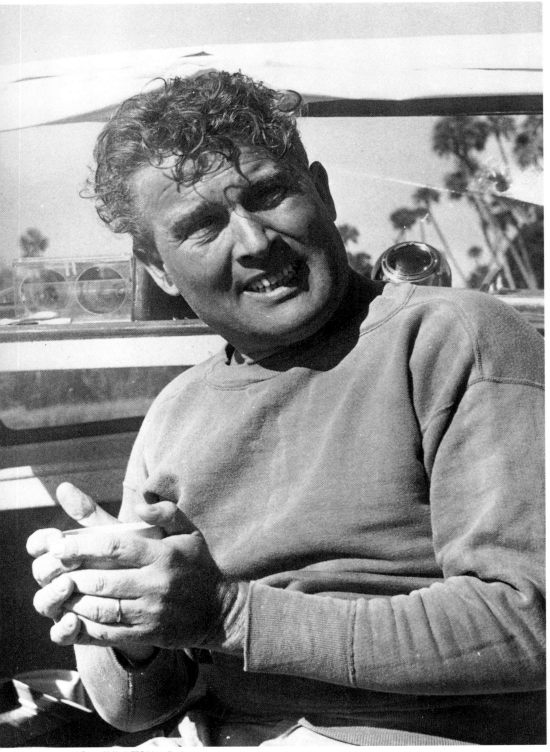

© Burton McNeely/FPG International

For forty-three years NACA excelled in carrying out its chartered mandate: "To supervise and direct the scientific study of the problems of flight, with a view of their practical solutions."

During 1917, the inventor Charles Kettering worked with the Delco and Sperry companies experimenting with what is considered by most to be the first United States guided missile, a pilotless biplane known as the *Bug*.

Made largely of wood, the small plane weighed only 600 pounds (273 kilograms), and 300 (136) of that was the bomb payload. It was powered by a forty horsepower Ford engine. Takeoff was accomplished from a four-wheel carriage running along a portable track. Flight direction was controlled by a small gyro, and altitude by an aneroid barometer.

Once target distance and wind conditions were determined, the number of revolutions required by the engine for the *Bug* to arrive at its target was calculated and a cam (a projecting part of a wheel used to give an alternating or variable motion to another wheel or piece) was set. Once the engine sent the missile the required distance, the cam dropped into position. The bolts that fastened the wings to the fuselage were pulled in, detaching the wings and dropping the missile onto its target. The *Bug* was tested successfully in 1918 before a gathering of Army Air Corps observers in Dayton, Ohio.

The success of the *Bug* was due to the gradual acceptance of the idea that the rocket was the key to space travel. While it was being tested, the potential of the rocket was realized independently by three different men, born into diverse cultures. Konstantin E. Tsiolkovsky of Russia conducted the first scientific study of rocket propulsion for space vehicles. Robert H. Goddard of the United States not only recognized that the entire science of astronautics rested on the rocket propulsion system, but also successfully tested the world's first liquid-fuel rocket in Auburn, Massachusetts on March 16, 1926. (The rocket, which stood ten feet [three meters] tall, accelerated to a speed of sixty mph [ninety-seven kph] and flew 184 feet [56.08 meters].) And the German, Hermann Oberth who, in the early 1900s, theorized that a propulsion system could operate in the near-vacuum of outer space. These men all came to the same conclusions about

Dr. Wernher von Braun (*opposite page, above*) helped to shape the systems approach that was necessary to achieve space flight. Dr. von Braun was recruited from Germany by United States Intelligence. *Left:* A photo of Dr. Robert Hutchings Goddard, the "Father of American Rocketry" and one of the pioneers in the theoretical exploration of space. Note his shop in the background. Criticized throughout his lifetime, Goddard's work was finally accepted when it was discovered that it was virtually impossible to build and launch a rocket without using some of the devices he developed. The photo on the opposite page, below, shows how Dr. Goddard transported his rockets to his launch tower, located fifteen miles (twenty-four kilometers) northwest of Roswell, New Mexico.

the future of space travel: Their conclusions, that rockets would carry man to space, became the basic working formulas of our Space Age.

In June 1920, the first NACA laboratory, the Langley Memorial Aeronautical Lab, in Hampton, Virginia, was dedicated. Here, aeronautics was the major research effort, utilizing wind tunnels as the chief tools. By 1930, the results of this research were so impressive that the facility had gained worldwide recognition. The up-to-date wind tunnels were hailed as far-seeing. The NACA Cowling (1928)—a streamlined design for aircraft engines—developed here for air-cooled radial engines, utilized a streamlined shape that increased aircraft speed and led to the low-wing, multiengined air transports and bombers of the 1930s.

Systematic studies of aerodynamic drag reduction led to improved design practices, including the development of retractable landing wheels over fixed, and therefore exposed, landing gear.

A second research center, the Ames Aeronautical Laboratory, was constructed near San Francisco in 1939 with a wind tunnel that dwarfed its predecessor at Langley. A third facility, later named the Lewis Flight Propulsion Lab, was built in Cleveland, Ohio, in 1940 to perform basic research, develop and test aircraft engines and study fuels. Research on the jet engine began there in 1943.

With the advent of the Second World War and an increased demand for combat planes, the focus of research and development turned to aerodynamics and structure. This refocusing resulted in extremely effective fighter planes. Although great strides were made in the development of the rocket, advances awaited modern technology and the team effort of thousands of minds simultaneously working on hundreds of complex problems. Under the leadership of Major General Walter R. Dornberger and Dr. Wernher von Braun (both of whom were later recruited by United States Intelligence to work for its rocket programs), the German test station at Peenemünde opened in 1937, signaling the dawn of the systems approach that was necessary to achieve space flight.

GODDARD—THE FATHER OF MODERN AERONAUTICS

During the years of aeronautic progress, a lone figure, Dr. Robert H. Goddard, had been investigating rocket flight. In 1919, he published a paper titled "A Method of Reaching Extreme Altitudes."

As an outgrowth of this and other theoretical work, on March 16, 1926, he successfully tested and launched the first liquid propellant rocket—despite the fact that journalists often misunderstood him and referred to him in print as "the moon man."

For the next two decades, Goddard conducted research, built and flew rockets, provided a mathematical analysis for multistage rockets and, by the time of his death in 1945, had amassed more than 150 patents.

By the end of World War II, NACA's research had led to rocket propulsion; finally, air and space flight met. This led to the development of the so-called X series of rocket research aircraft in 1944. The X-1 was built specifically to investigate the transonic region and to break the sound barrier. On October 14, 1947, Air Force Captain Charles E. "Chuck" Yeager piloted the X-1's historic barrier-breaking flight.

The most famous of the X research planes was the X-15. Still an idea on a drawing board in 1952, the X-15 finally achieved its anticipated altitude and speed objectives in 1968, spanning the transition from aeronautical research to the new Space Age.

Postwar research into higher flight speeds led to high-altitude drop-test models for gathering flight data, then to the use of rockets to launch models to transonic (from just below to just above the speed of sound) and supersonic speeds.

Langley acquired a surplus naval station on Wallops Island, Virginia, and named it the Pilotless Aircraft Research Division. Next, a High-Speed Flight Research Station was established at Muroc, California (later, this would become the famous Edwards Air Force Base), for a series of specific research aircraft.

At Langley, a transonic wind tunnel was developed and built in 1950; researcher Richard T. Whitcomb used this tool to discover the "area rule" (the cross section areas of an aircraft

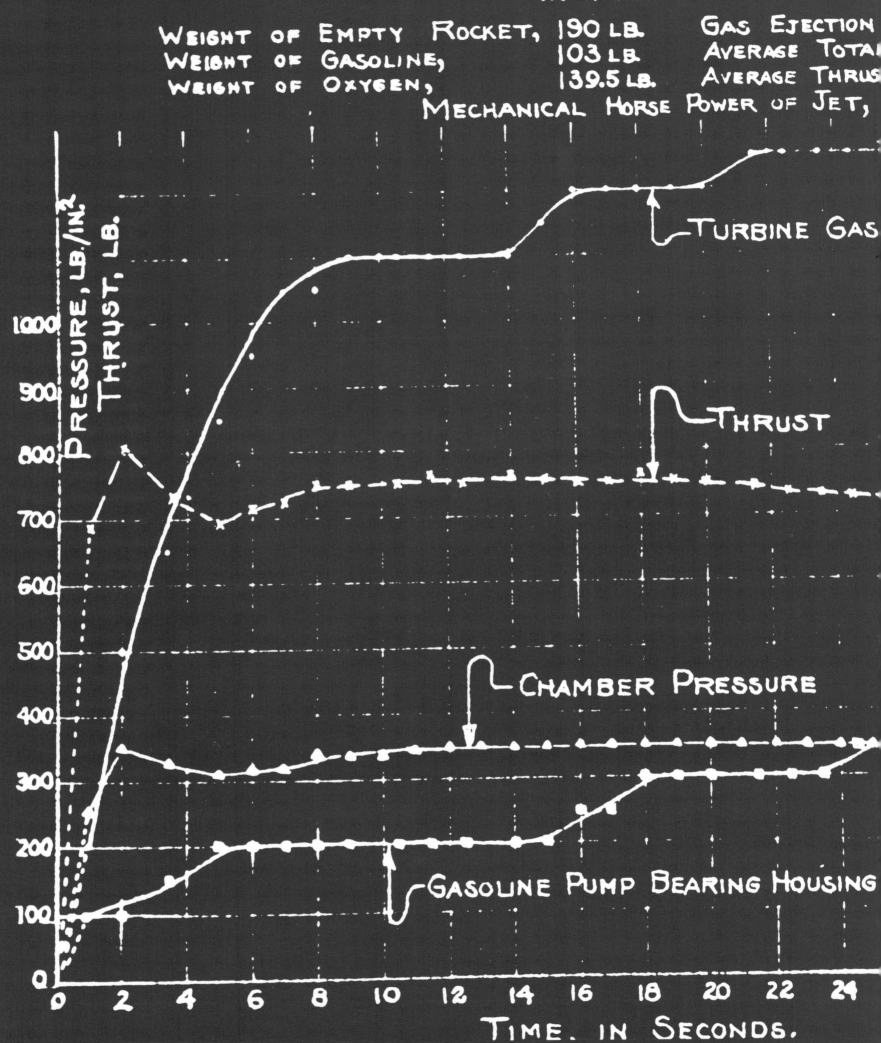

DATA FROM TEST No
IN PUMP SERIES

WEIGHT OF EMPTY ROCKET, 190 LB. GAS EJECTION
WEIGHT OF GASOLINE,   103 LB. AVERAGE TOTAL
WEIGHT OF OXYGEN,   139.5 LB. AVERAGE THRUS
MECHANICAL HORSE POWER OF JET,

TURBINE GAS

THRUST

CHAMBER PRESSURE

GASOLINE PUMP BEARING HOUSING

PRESSURE, LB./IN.²
THRUST, LB.

1000
900
800
700
600
500
400
300
200
100
0

0 2 4 6 8 10 12 14 16 18 20 22 24

TIME. IN SECONDS.

4

LOCITY, 4,000 FT./SEC.
ATE OF FLOW, 6.0 LB./SEC.
ER LB. TOTAL FLOW PER SEC., 12.5 LB.
20 H.P.

EMP

260

EMP.

20

28 30 32 34 36 38 40

TITLE: Rocket Test Results

DR.:
CH.:

ED:

GUGGENHEIM ROCKET RESEARCH PROJECT
DR. R. H. GODDARD, DIRECTOR, P. O. BOX 978, ROSWELL, N. MEX.

The test results of Dr. Goddard's rocket launch of June 24, 1941 are shown here. *Inset:* A Vanguard TV (Test Vehicle) being launched on October 23, 1957. Fourteen *Vanguard* rockets were tested at the Atlantic Missile Range, Cape Canaveral, Florida. These paved the way for the launch of *Vanguard I,* on March 17, 1958.

Pilot Steve "Chuck" Yeager flew this "X-series" plane.
Yeager was the first pilot to break the sound barrier.

Right: Chuck Yeager (on right) standing next to his plane, an X-11. *Opposite page, above:* This cut-away of the Soviet earth satellite illustrates its protective shield. Once the satellite enters orbit, it is tossed away by means of a spring located at the top of the craft. Also visible are the scientific research instruments and a hermetically sealed cabin to house an experimental animal (such as Latka). *Opposite page, below:* A model of *Explorer 1* in a simulated space atmosphere. *Explorer 1* was placed in orbit on January 31, 1958 by a Jupiter-C launch rocket. This eighteen pound (eight kilogram) satellite, the first launched by NASA, discovered the first of two circular radiation belts (later named the Van Allen radiation belts) that surround the earth.

© PHOTOWORLD/FPG International

should not be altered too rapidly from the front to the back of the plane). A genuine breakthrough in airplane design, the tunnel's immediate application allowed military aircraft to break the sound barrier while in level flight.

From July 1957 to December 1958, the International Geophysical Year (IGY) was observed. Science academies of sixty-seven nations had agreed to designate the period of IGY and to pool their national scientific resources in a coordinated data gathering program aimed at studying the Earth as a planet. Its scientific program included a proposal to launch satellites that would measure Earth from space.

During this year the question of what type of rocket should be used to launch the first United States satellite arose. Three options were discussed by a specially appointed committee: the Air Force *Atlas* rocket, the Army *Redstone* and a brand-new Navy rocket called the *Vanguard*.

After long deliberation, the committee selected the Navy's *Vanguard* rocket, their rationale being that since

the IGY was a peaceful, international science project, it would be inappropriate to use well-tested military hardware such as the *Atlas* or the *Redstone*.

The Soviet Union, however, did not feel the same way and used its flight-tested, twenty-engine rocket to launch *Sputnik* on October 4, 1957. The space race was indeed on. As a direct result, the United States' Army launched the *Explorer 1* satellite—so named because its mission was to explore the unknown—four months later.

Explorer 1 fulfilled the United States' commitment to IGY, and its small packages of on-board measuring instruments produced the first major discovery of the Space Age, the Van Allen radiation belts surrounding Earth.

As a result of the space race, the United States successfully began launching satellite after satellite, and both the *Vanguard* and the new *Atlas* proved they could do the job. The success of *Sputnik* also spurred the creation of a new agency to develop a national space program, with which President Eisenhower wanted to

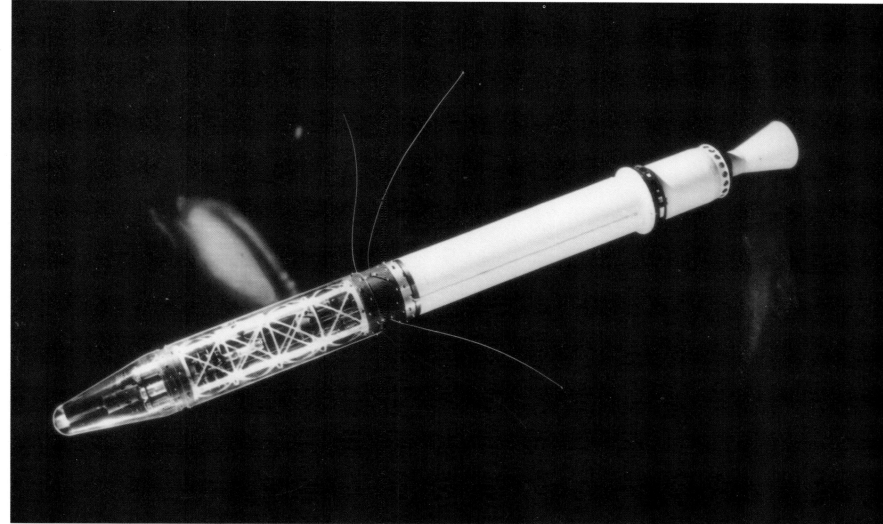

Opposite page, right to left: Dr. Wernher von Braun, Dr. James A. Van Allen and Dr. William H. Pickering hoist a model of *Explorer 1* after its launching on January 31, 1958. Dr. Van Allen's scientific experiments discovered the radiation belts that surround the Earth. *Left:* Placed in orbit on September 18, 1959, the *Vanguard III* satellite was instrumental in gathering and transmitting data on the Earth's magnetic field, solar x-rays and micrometeoroids. The success of this satellite completed the Vanguard experiments, the United States' first scientific satellite program. *Below:* This artist's conception shows how *Mariner* looked as it flew by the planet Venus. Its mission was to take infrared and microwave measurements of the planet, communicate this information to earth over an interplanetary distance of thirty-six million miles (57.6 million kilometers) and obtain data on interplanetary phenomena during its trip to Venus.

A photograph of a K-15 airplane dropping away from a B-52 bomber in flight.

emphasize the peaceful uses of research and development. Three agencies vied for leadership: the Atomic Energy Commission, the Department of Defense and NACA.

The NACA proposal combined aeronautics and space research with a solid scientific base. The committee also offered experience in working closely with the military as well as providing research for civil applications. By April 1958, the administration's position and the NACA proposal had been combined into a single bill for creating a national aeronautics and space agency. And so, on July 29, President

Eisenhower signed the National Aeronautics and Space Act of 1958, and NASA was born.

With NACA as its nucleus, the NACA staff, facilities, programs and responsibilities were transferred over to the civilian agency, NASA. The prior organization's tradition of excellence became a legacy to the new organization when NACA ceased to exist on September 30, 1958.

On October 1, NASA came into being, "devoted to peaceful purpose for the benefit of all mankind." The United States had finally come of age in the Space Age.

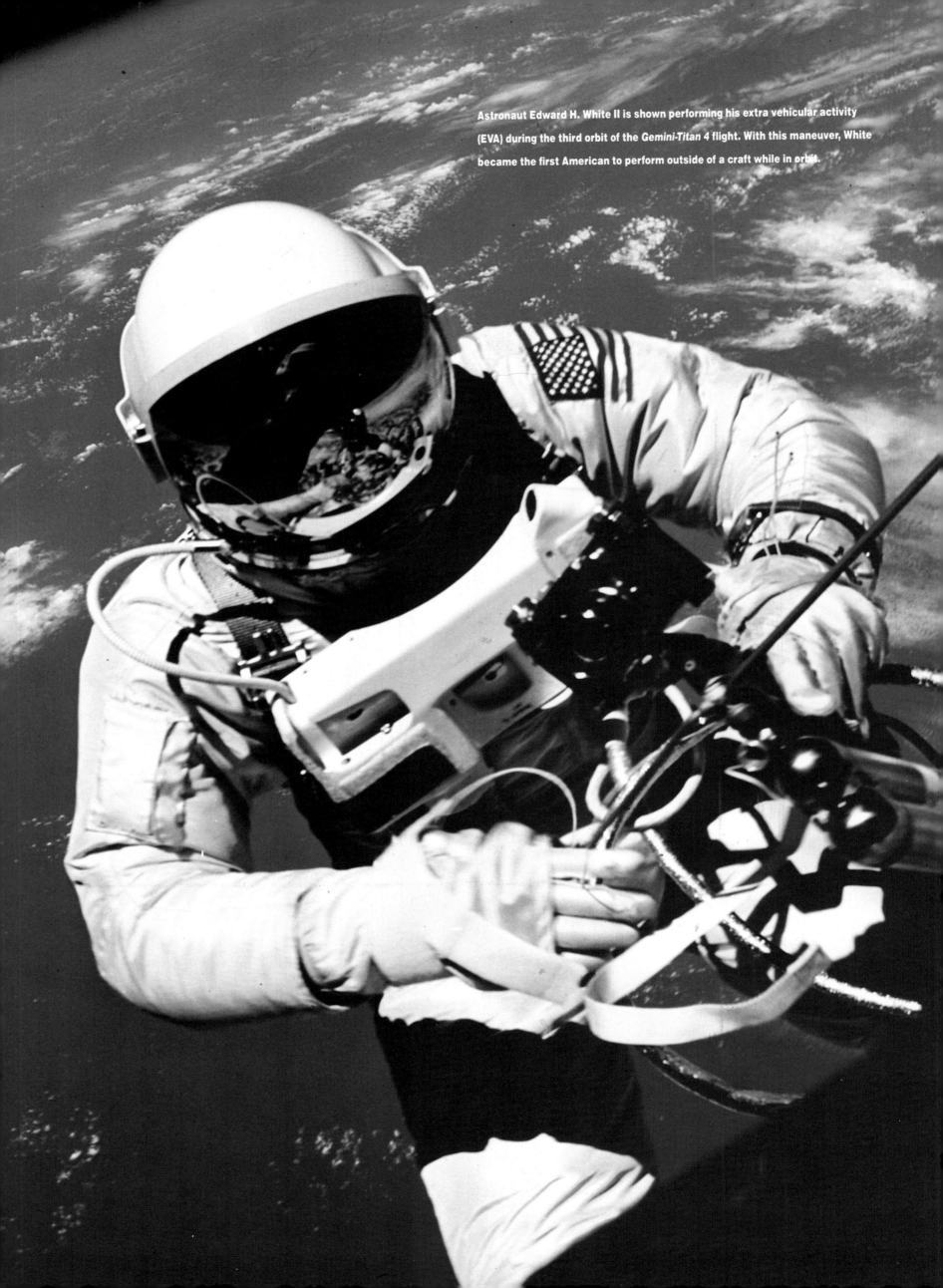

Astronaut Edward H. White II is shown performing his extra vehicular activity (EVA) during the third orbit of the *Gemini-Titan* 4 flight. With this maneuver, White became the first American to perform outside of a craft while in orbit.

CHAPTER 2

"There are only two ways of learning to ride a fractious horse; one is to get on him and learn by actual practice how each motion and trick may be best met; the other is to sit on a fence and watch the beast awhile, and then retire to the house and at leisure figure out the best way of overcoming his jumps and kicks. The latter system is the safer, but the former, on the whole, turns out the larger proportion of good riders. It is very much the same in learning to ride a flying machine; if you are looking for perfect safety you will do well to sit on a fence and watch the birds, but if you really wish to learn you must mount a machine and become acquainted with its tricks by actual trials."

—WILBUR WRIGHT

LEARNING TO RIDE

By the late 1940s it was clear that the Army's burgeoning missile program was outgrowing its Fort Bliss test site in El Paso, Texas. So NACA moved seven top-notch scientists (including Wernher von Braun) to a new facility in Huntsville, Alabama. There, they developed the *Hermes* vehicle, which was not extremely successful but opened the door for the development of the *Redstone* missile—the first major American effort in the field of rocketry—and, later, NASA's first rocket designed specifically to put an American in space.

A new chapter in rocket history was created when the decision to expand the Redstone Arsenal was reached. After a lengthy search,

A Jupiter-C rocket beginning to lift-off. This rocket, with its tremendous lift capability, was used to launch many NASA crafts.

more than 500 military personnel, 130 members of the original von Braun team, several hundred General Electric employees and 120 civilian government workers settled into the modern, new facilities.

The Redstone organization was designated as the Ordnance Guided Missile Center (OGMC). It was headed by Major James P. Hamill, who had been transferred from the Rocket Branch of the Ordnance Research and Development Division of the Pentagon. Major Hamill and the vanguard of the group had hardly settled in when the Korean conflict, as it was called, broke out in June 1950.

The OGMC's first assignment was to conduct a feasibility study for a 500-mile (804.65-kilometer), ballistic surface-to-surface missile. As the war in Korea grew more intense, the missile's priority increased. After being unofficially called "Ursa" and, later, "Major," the project was baptized *Redstone*, after the arsenal where it was being developed, on April 8, 1952.

Deciding against creating an entirely new engine, the group used a modification of the liquid propellant engine developed earlier by North American Aviation for the Navaho missile, the precursor of the Intercontinental Ballistic Missile (ICBM). The Navaho program was instituted by the Air Force in 1951 with the advent of its *Atlas* missile.

As the war and the program proceeded, the Army's requirements changed, reducing the missile's desired range from 500 miles to 200 miles (806 to 322 kilometers). This reduction provided a bonus: The sturdy *Redstone* would be able to carry a nuclear warhead and would, if necessary, become a mobile weapon capable of being launched under battlefield conditions by United States combat troops.

The first *Redstone* was fired with moderate success from Cape Canaveral, Florida, on August 20, 1953. It traveled an 8,000-yard (7,315-meter) trajectory. Thirty-six more research and development (R&D) models were launched through 1958, sixteen of them built by the Redstone Arsenal, the rest by the Chrysler Corporation.

The years between 1952 and 1954, during which the *Redstone* was designed and developed, were critical ones in the history of the United States

missile program, as the basis for today's launch rockets was established during this period. Another crucial development was the growing realization that the Soviet Union was threatening a breakthrough in weaponry that appeared to endanger the very existence of the United States.

The Soviets, who produced their first atomic bomb in 1949, ended the United States monopoly on the hydrogen bomb when they exploded their own on August 12, 1953. Before long, military intelligence produced evidence that the Russians were rapidly developing a ballistic missile program and, furthermore, might be well ahead of the American effort.

This information fueled all three branches of the armed services. Immediately, they began developing their own vigorous programs to produce long-range ballistic missiles, each working with its specialty: the Army on short- and intermediate-range missiles, the Navy on missiles to be fired from seagoing vessels and the Air Force on its long-range cruise missile program and the new *Atlas* ICBM.

The Army's contribution to this program, which came into existence in February 1956, developed the modified *Redstone* rocket, called the *Jupiter A.* Between September 1955 and June 1958, twenty-five *Jupiter A*'s were fired.

Jupiter C, officially called the Jupiter Composite Reentry Test Vehicle, was designed primarily to test nose cone materials for the forthcoming IRBM (Intermediate Range Ballistic Missile). In addition, it provided the capability for launching a satellite into orbit. It was later known as the *Juno 1* vehicle.

In its first flight, on September 20, 1956, *Jupiter C* reached the unprecedented altitude of 682 miles (1097.5 kilometers), and landed 3,400 miles (5,471 kilometers) from Cape Canaveral—a record unequalled by the United States for two years.

The second launching in the summer of 1957, lifted a scaled-down *Jupiter* nose cone into space, but guidance difficulties caused it to land outside of the target area. The third and final test, on August 8, 1957, resulted in the retrieval of the revolutionary nose cone after it flew a trajectory 300 miles (482.8

kilometers) high, and 1,200 miles (1931.1 kilometers) long.

In 1956, as work on the *Jupiter* continued, advances in thermonuclear technology (the A-Bomb) caused the Army to lose out in the interservice rivalry for control of the nation's IRBM-ICBM program. At this time, President Eisenhower relieved the Army of control of the program; he first switched responsibility for the program to the Department of Defense and then later, to the Air Force.

IN SPACE FOR PEACE

While *Jupiter* progressed as an Air Force project, it was a non-military *Jupiter* flight that sounded a note for the future. On May 28, 1959, two monkeys named Able and Baker rode a *Jupiter* rocket 300 miles (482.8 kilometers) high and 1,600 miles (2574.9 kilometers) downrange and were recovered alive. The forerunners of other living space payloads, they were the first of several "monkeynauts."

Only after several months of grueling tests were the chimps allowed into the early spacecrafts, and even then they were connected at their hands and feet by electrodes and periodically given small electric shocks in case they made incorrect moves.

T. Keith Glennan, who had been president of Case Western Institute of Technology in Cleveland, Ohio became the first NASA administrator in 1959. He stressed the strong need for a group of scientists and engineers with experience in the fields of missiles and carrier rockets and urged them to become a part of NASA.

The only group capable of carrying out Glennan's wishes was the Medaries-von Braun team at the Redstone Arsenal. On July 1, 1960, the Huntsville facility (the Redstone Arsenal) was personally dedicated by President Eisenhower as the George C. Marshall Space Flight Center, with Wernher von Braun serving as its director. Congress immediately approved the $915 million budget requested by NASA for the coming fiscal year.

A large part of that budget was slated

for a carrier that had been conceived at the Huntsville facility before NASA was founded. In 1956, the Development Operations Division at Redstone had begun studies on carriers that would go far beyond anything else that was currently being planned, researched or developed.

Von Braun knew that the ICBM-based carriers would be inadequate not only for manned missions in Earth orbit, but also for future manned lunar and interplanetary flights. Thus, the vehicle and propulsion engineers at the Army Ballistic Missile Agency (ABMA) set out to see how they could develop the basic elements of the *Jupiter* and *Redstone* rockets for use in Earth, lunar and interplanetary orbit.

The idea was to cluster several *Jupiter* engines around *Redstone* and *Jupiter* propellant tanks, building a larger carrier vehicle that could fully utilize the experience, the hardware and the facilities of ABMA, as well as its associated Army and industrial supporter. From this emerged the *Saturn* rocket.

By October 1959, the Department of Defense's Advanced Research Project Agency (ARPA) had developed four *Saturn* designs. Each had essentially the same first stage, with distinctive variations in the upper stages. The NASA-Defense Saturn Vehicle Evaluation Committee picked one of the four configurations, the *Saturn C-1*, and began developing it in December. Because there was close cooperation between NASA and ARPA on the *Saturn* project from its outset, it was simple and logical to transfer first *Saturn*, then the entire von Braun science team, to the civilian space agency, NASA.

On March 3, 1959, the first man-made object was launched from Cape Canaveral into a trajectory that carried it forever out of Earth's gravitational field.

The spacecraft itself, *Pioneer 4,* had a minimum of scientific instrumentation, but its telemeter transmitter provided irrefutable proof that it had shot past the Moon at a speed sufficient to escape Earth's gravity. The launch rocket, *Juno 2,* was a make-do configuration from the Army's stable: The three-stage cluster of small, solid rockets of the successful *Juno 1* was borrowed from the *Redstone* and placed into the nose of the more powerful *Jupiter* IRBM.

The upper stages of the *Saturn C-1*

(later simply called *Saturn 1*) configuration were approved at the end of 1959; after that, vehicle research, development and test-flight programs began.

The second-stage, designated S-4, would have four 20,000-pound (9,072 kilograms) thrust engines; the third stage, S-5, had two such engines. After *Saturn* won congressional approval, it was given the coveted "DX" rating, meaning it had high priority for materials, personnel and other resources.

The United States, following a series of discouraging failures in 1962, scored outstanding successes in 1964 and 1965 with *Rangers 7, 8* and *9,* which transmitted back to Earth thousands of pictures of craters as small as a few feet across on the lunar surface. *Ranger* was followed by *Surveyor* and the *Lunar Orbiter* series, which provided vital data for the six *Apollos* that would land on the Moon between the years 1969 and 1972.

Below: This smiling monkey, called a pigtail monkey, was the type used and trained for early biosatellite flights at NASA's Ames Research Center at Mountain View, California. Some of these monkeys were in space for up to thirty days to test the effects of weightlessness on the nervous and cardiovascular systems. The equipment shown in this photo is test hardware and is not flight-rated. The photograph on the opposite page is of the first astronaut team; their selection was announced on April 9, 1959, only six months after the formation of NASA. They are, front row, left to right: Walter M. Schirra Jr., Donald K. Slayton, John H. Glenn Jr., and Scott Carpenter; back row left to right: Alan B. Shepard Jr., Virgil I. "Gus" Grissom and L. Gordon Cooper.

FURTHER TOWARD THE UNKNOWN

Meanwhile, in 1959, after sending several chimp crews into space, NASA chose the seven men who would become America's first astronauts. Despite taunts from Edwards Air Force Base personnel (all seven men chosen were from the Air Force) that they were going to "follow monkeys into space," the men became instant heroes.

The physical and emotional rigors of traveling in space required that NASA pick strong-minded men in top physical condition. The seven men chosen were Alan Shepard Jr., John Glenn, Scott Carpenter, Walter Schirra, Gus Grissom, Donald Slayton and L. Gordon Cooper. Together these men had logged a total of over 2,000 hours of flight time in what were, at that time, state-of-the-art jets.

These astronauts-to-be were put through every possible psychological and physical test that NASA scientists could dream up: claustrophobia tests, treadmill tests, the centrifuge test (for the effects of and lack of gravity) and even a tour of duty in the jungles of Central America. After their selection, a leader was chosen, the first man who would venture into the mystery of space. By all accounts, this selection amounted to nothing more than a personality contest. Alan B. Shepard, with his strong character and impressive military record, was chosen as the first astronaut to serve.

Right: This huge motor was built for NASA by the Aerojet-General Corporation near Miami, Florida. It is a 260-inch (780-centimeter) diameter, solid fuel rocket motor which developed 3.5 million pounds (1.6 million kilograms) of thrust. Enclosed in a special steel casing, the motor contained 1,680,000 pounds (763,636 kilograms) of solid propellant cast in one piece. *Below:* Before being connected to a *Delta* launch vehicle, this S-3a energetic particle satellite underwent inspection of its antennas. This satellite is similar to the *Explorer XIII* but had improved features, such as the capability to take refined measurements of energetic particles and to study their relationships to the magnetic fields of earth and interplanetary space. *Opposite page:* At Cape Canaveral a technician assembles the arm from which the *Mariner II* high-gain antenna was suspended. This antenna permitted *Mariner II* to transmit information gathered from the vicinity of Venus to Earth, thirty-six million miles (57.6 million kilometers) away.

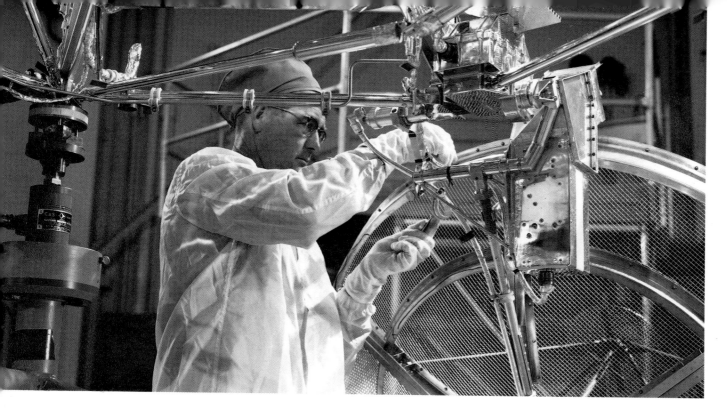

The inauguration of John F. Kennedy as president of the United States brought about many changes in the space program. One of these changes was being made to *Saturn's* upper stages. Six 15,000-pound (6,803-kilogram) thrust engines, all using the high-energy combination of liquid oxygen and liquid hydrogen, replaced the four 20,000-pound (9,072 kilogram) thrust second-stage engines. This 15,000-pound engine was identical to the 20,000-pound engine except for the reduction in its thrust rating. The third stage was completely eliminated. On April 29, 1960, all eight engines were ignited for eight seconds, producing 1.3 million pounds (589,676 kilograms) of thrust, a NASA record.

At Huntsville and at Cape Canaveral, new facilities had been built. The assembly of the first *Saturn* vehicle began in May 1960. Meanwhile, the Rocketdyne division of North American Aviation Inc. and the Pratt and Whitney Aircraft division of the United Aircraft Corporation continued to work on improving the *Saturn* engine.

In 1961, President Kennedy passionately committed the United States to "land a man on the Moon before the decade is out."

The first decision NASA had to make was how, exactly, the lunar flight mission was to be executed. Three conceivable plans were studied in great detail:

1) The direct mode, where a single, very powerful rocket would launch a heavy spacecraft capable of a direct soft-landing on the Moon and of subsequently reascending. (This mode would have required a first stage with eight rockets, instead of *Saturn 5's* five-rocket engine.)

2) The Earth-orbit rendezvous mode, where the launch rocket's third stage—the stage that must drive the same heavy spacecraft of Mode 1 from low Earth orbit to translunar injection—would be refueled in Earth orbit by propellants carried by a second, identical, launch vehicle. (It was discovered that with this concept the total boost load could be distributed between two smaller, five-engine *Saturn 5's*.)

3) The lunar orbit rendezvous mode, where a single launch vehicle (NASA hoped that a single five-engine *Saturn 5* would work) would launch a spacecraft configuration consisting of three modules directly into a translunar trajectory. The propulsion system of the service module would first induce itself, then the two other modules, into an orbit around the Moon. The command module, in which the astronauts would ultimately return to Earth, would remain in lunar orbit, still attached to the service module. Meanwhile, a lunar module, solely designed for the lunar surface portion of the flight, would descend from lunar orbit to the Moon's surface and later reascend to the orbit for a rendezvous with the command and service module. The latter two modules would return to Earth, and only the heat-protected command module would survive the blazing reentry into and through the atmosphere. This mode was ultimately selected.

During 1961, *Saturn 1's* role became clearly defined as a test vehicle for the *Apollo* program. Two editions of the carrier would be built, Block 1 and Block 2.

The Block 1 *Saturn*s would have

dummy, or mock, upper stages and would be fired to test the basic concept of the vehicle; Block 2's would carry more propellants and would be powered by upgraded engines, which would develop 188,000 pounds (85,276 kilograms) of thrust each. The Block 2 *Saturn*s would have stabilizing tail-and-stub fins; the Block 1's would not. Later carriers would also have live S-4 upper stages, an improved instrument unit and a boilerplate, or dummy, model of the *Apollo* capsule.

After the first *Saturn*, designated *SA-1*, was static tested (the rocket was fired without launching it) in Huntsville in May 1961, plans were made to transfer it to Cape Canaveral. On October 27, *SA-1*, a 162-foot-long (49.4 meters) carrier, weighing nearly one million pounds (453,597 kilograms), majestically lifted off the ground in a virtually flawless maiden flight. As it flew its short 200-mile (322 kilometer) trajectory, more than 500 different measurements were recorded, including flight pattern, fuel consumption, vehicle stability and thrust distribution.

In 1962 and 1963 the rest of the Block 1 vehicles were fired without a hitch. The first Block 2 vehicle was launched on January 29, 1964. Its second stage propelled a total weight of 37,700 pounds (17,101 kilograms) of payload into orbit.

Dummy *Apollo* capsules were orbited in May and September by *Saturn SA-6* and *SA-7*; these flights proved that the spacecraft and its carrier were compatible. The final three Block 2 *Saturn*s orbited *Pegasus* micro-meteoroid-detection satellites, as the *Saturn 1* program ended with a 100 percent successful flight test record.

Meanwhile, on April 12, 1961, Soviet Cosmonaut Yuri Gagarin, launched by a rocket the Soviets referred to as *A-1*, was orbiting Earth in his 10,395-pound (4,715-kilogram) *Vostok* space capsule. The United States immediately responded with its first astronaut, Alan B. Shepard Jr., and his 200-mile (322.9-kilometer) suborbital ride on May 5, 1961, in a *Mercury 3* rocket inside his capsule, *Freedom 7*. The *Mercury* rocket was launched by the *Redstone* launch vehicle.

The initial space capsule designed by NASA was, at best, uncomfortable. The capsule held only enough room to lower the single astronaut in with all the

May 5, 1961. A United States Marine Corps helicopter recovery team poised above astronaut Alan B. Shepard Jr. after his successful suborbital flight, the first in the Project Mercury Program.

necessary instrument panels, radio hook-ups, tanks, tubes and the emergency parachute. The astronaut's seat was sculpted to his back and he could only view the outside world through a periscope. Instead of a hatch, there were two tiny portholes above the right and left sides of his head.

At the beginning of the *Mercury* launches, NASA officials did not see the need for the astronaut to control the craft; However, they were persuaded by the astronauts themselves that more control within the craft itself was an absolute necessity.

By the second *Mercury* launch, NASA had redesigned the capsule, creating a pilot's window as well as a hatch that the astronaut could open himself upon touchdown.

On August 6, 1961, the Soviets bettered Gagarin's two-orbit flight with a seventeen-orbit, twenty-five-hour flight by cosmonaut Gherman Titov.

In response, on February 20, 1962, America demonstrated that it, too, was capable of, and ready for, manned orbital flight. John Glenn was launched by a massive *Atlas D* vehicle in the *Mercury 6*, in his capsule, *Friendship 7*, thus becoming the first American to orbit the earth. Glenn completed three full orbits. Soon after, on May 24 of the same year, Scott Carpenter was hurled into space in an almost identical flight.

It was with Carpenter's flight that NASA received its first major scare. While Glenn's flight had been strictly business, Carpenter was more science-minded and was given experiments to perform by NASA's life scientists. As he twirled about in space, locating glittering dust particles over Mexico, his fuel began to run out. NASA ordered him to return immediately, but there was no signal from his craft—Carpenter had vanished!

As Walter Cronkite began to announce on the CBS Evening News that NASA had lost its first astronaut, Carpenter's craft burst through the atmosphere and plunged into the Atlantic Ocean, miles away from his anticipated splashdown site at Cape Canaveral.

In 1962, the Air Force increased the payload capability of its *Titan 2* ICBM; the new configuration was known as *Titan 3C*.

The *Titan 3C* was the first United

Opposite page, above: Astronaut John Glenn explaining the intricacies of the Astronaut Space Glove to President John F. Kennedy during the President's tour of United States space installations. Opposite page, below: Astronaut "Wally" Schirra, prior to his Mercury mission. Below: Back-up astronaut Gordon Cooper assists astronaut Walter Schirra into his Sigma 7 spacecraft. The Atlas 8 was successfully launched from Cape Canaveral, Florida, on October 3, 1962. Left: Traveling at 17,500 mph (28,000 kph) and in a state of weightlessness, astronaut John Glenn was photographed in space by an automatic sequence motion picture camera.

States carrier to use both parallel and tandem staging. Its 120-inch (304.8-centimeter) diameter solid fuel boosters could be replaced by a 156-inch (396.2-centimeter) booster for even greater lifting capability. Since the third stage could be restarted in space, the *Titan 3C* could change the orbit, or inclination, of its spacecraft up to 25,000 miles (40,322 kilometers) from Earth, even sending them on interplanetary trajectories. On its first flight from Cape Kennedy (formerly Cape Canaveral), *Titan 3C* placed a 21,000-pound (46,297 kilogram) payload in a nearly circular orbit over 100 miles (161 kilometers) high.

On October 3, 1962, Commander Walter Schirra Jr., completed six orbits around the Earth in the *Mercury 8* spacecraft, *Sigma 7*. During May 15 and 16 of 1963, Major L. Gordon Cooper Jr., in the *Mercury* capsule *Faith 7*, completed a total of twenty-two Earth orbits.

During an orientation session at the McDonnell Aircraft Corporation plant in St. Louis, Missouri, astronauts Virgil Grissom and Neil Armstrong explore the two-man *Gemini* spacecraft. McDonnell was the prime contractor for the *Gemini*, the follow-up program to Project Mercury. Standing on the floor are astronaut Charles Conrad and the Vice President of McDonnell Aircraft Corporation, Walter Burke (in glasses) discussing the mock-up's details.

THE GEMINI MISSIONS

A total of six *Mercury* spacecrafts were launched by NASA from 1961 through 1963. Through these flights, the astronauts demonstrated their ability to control their craft manually while weightless. On March 23, 1965, the single-seat Mercury flights gave way to the two-seat Gemini capsule.

NASA's goal with *Gemini* was, primarily, to orbit men in space for at least the seven days it would take the *Apollo* to reach the Moon, land and return. Because the *Gemini* was a larger and heavier rocket, a more powerful launch rocket was needed. NASA immediately called on the Air Force's *Titan II*.

From June 3 through June 7, 1965, NASA launched Major L. Gordon Cooper Jr. and Lieutenant Charles Conrad Jr. in the *Gemini 5*, making a total of 120 Earth orbits.

In August, aboard the *Gemini 4*, Edward H. White became the first astronaut to walk in space. Attached to a thirty-two-foot (eight-meter) "umbilical cord," White propelled himself via a hand-held gas gun. Over a period of nearly thirty minutes he took dozens of photographs from his unique vantage point.

The *Gemini 6* spacecraft, which was to have been the next in orbit, was postponed from launch in October 1965 because the *Agena*, with which it would practice rendezvousing, was destroyed after launch. Instead, *Gemini 7* was sent up first, on December 4, 1965, and the same launch pad was quickly readied for *Gemini 6*. While Lieutenant Colonel Frank Borman and Commander James A. Lovell Jr. were orbiting in *Gemini 7*, Walter Schirra and Major Thomas P. Stafford went into orbit on December 15.

Gemini 6 caught up with its sister craft and, in a series of impressively precise and complex maneuvers, came within one foot (thirty centimeters) of *Gemini 7*, approximately 185 miles (298 kilometers) above Earth. The two spacecraft flew in close formation for nearly eight hours, in a dramatic demonstration of *Gemini*'s ability to rendezvous in space. *Gemini 6* splashed down in the Atlantic Ocean after twenty-six hours aloft, but *Gemini 7* continued on for a record-breaking

Agena rendezvous—the *Agena* Target Docking Vehicle seen from the *Gemini 8* spacecraft during the latter's approach for rendezvous and station upkeep. This particular photo was snapped by astronaut David Scott at an approximate distance of 210 feet (sixty-five meters). During its sixth orbit the *Gemini 8* mission was terminated, probably due to an electrical short circuit in one of its thrusters.

Opposite page: During training exercises conducted in the Gulf of Mexico, off Galveston, Texas, astronaut Edward H. White seals the hatches of the *Gemini* training craft with a locking lever retrieved from the nose of the craft by astronaut James McDivitt (in the water). *Left:* At the Whirlpool Corporation in St. Joe, Michigan, a technician drinks orange juice during a test of a *Gemini* capsule.

206 orbits, taking it more than four million miles (2.48 million kilometers) in two weeks. The flight gave the United States nearly 1,353 man-hours in space, compared to 507 man-hours for the Russians.

Gemini 8 got off to a great start on March 16, 1966, when it went into orbit shortly after the launch of an *Agena* target vehicle. Late in the fifth orbit, after having accomplished the first docking operation in space, the coupled *Gemini 8* and *Agena* began pitching and rolling, which persisted after astronauts Neil A. Armstrong and Major David. R. Scott detached their craft, the *Gemini 8*, from the *Agena*. Quite worried, NASA quickly aborted the mission and *Gemini 8* splashed down in the western Pacific Ocean instead of the Atlantic Ocean as planned.

On May 30, 1966, NASA launched *Surveyor 1* to the Moon as a prelude of things to come.

NASA had planned to launch its *Gemini 9* spacecraft on June 1, 1966, directly after an "augmented target docking adapter" was orbited by an *Atlas*; however, due to ground communications problems, the launching was postponed until June 3. Colonel Thomas P. Stafford was commanding, and Lieutenant Commander Eugene A. Cernan was piloting the craft. In many ways the *Gemini 9* mission was the most significant to date, since it underscored the difficulties man must undergo when working in a truly alien environment.

Cernan's activities were so intense that he lost about ten pounds during the flight and two pounds of water were found in his spacesuit after recovery! His respiration rate increased to thirty

Gemini 7 splashed down in the western Atlantic recovery area at 9:05 a.m. (EST) on December 18, 1965, concluding a record-breaking fourteen-day mission in space. Astronaut James A. Lovell Jr., pilot of the flight, is hoisted from the water by a recovery helicopter.

Below: Watching the *Gemini 5* lift-off with astronauts L. Gordon Cooper, Jr. and Charles Conrad, Jr. aboard, are Mrs. Gordon Cooper and her daughters. *Opposite page, above:* This photo of *Gemini 7* was taken on December 15, 1965 from the hatch window of the *Gemini 6* spacecraft during rendezvous and station-keeping maneuvers. The craft is approximately 160 miles (256 kilometers) above the Earth. *Opposite page, below:* Another photo of *Gemini 6* during rendezvous.

breaths per minute, compared to a normal rate of twelve to fifteen per minute.

Both astronauts suffered considerable fatigue during the flight because their work load was about four times greater than had been anticipated. For instance, in a rendezvous phase, they had to resort to hand calculations when the on-board computer malfunctioned. When they finally approached the target of the rendezvous, once as close as three feet (ninety-two centimeters), they saw that the clamshell-like shrouds of the cone were open but still fastened. Stafford likened it to "an angry alligator" as it tumbled slowly through space. Ground controllers made several unsuccessful attempts to dislodge the shrouds by radio commands.

During this flight, Cernan took a space walk that lasted more than three times as long as Edward White's

extravehicular activity (EVA) had on the *Gemini 4* mission in 1965. Cernan, however, had difficulty with his maneuvering unit, and he exerted himself so much in his attempts to maintain position at the rear of the vehicle's adapter section, that he exceeded the limits of his spacesuit's life support system. This resulted in the fogging of his helmet visor, which so disturbed him that he was unable to use the maneuvering unit he was trying to put into service. Cernan reported that his tether (the umbilical-like cord that attached him to the ship) was of little aid in maneuvering. But he did accomplish many tasks, including making his way to the bow of the spacecraft, attaching a mirror there and evaluating the usefulness of Velcro patches to adhere to surfaces during zero-gravity flight.

The mission ended successfully

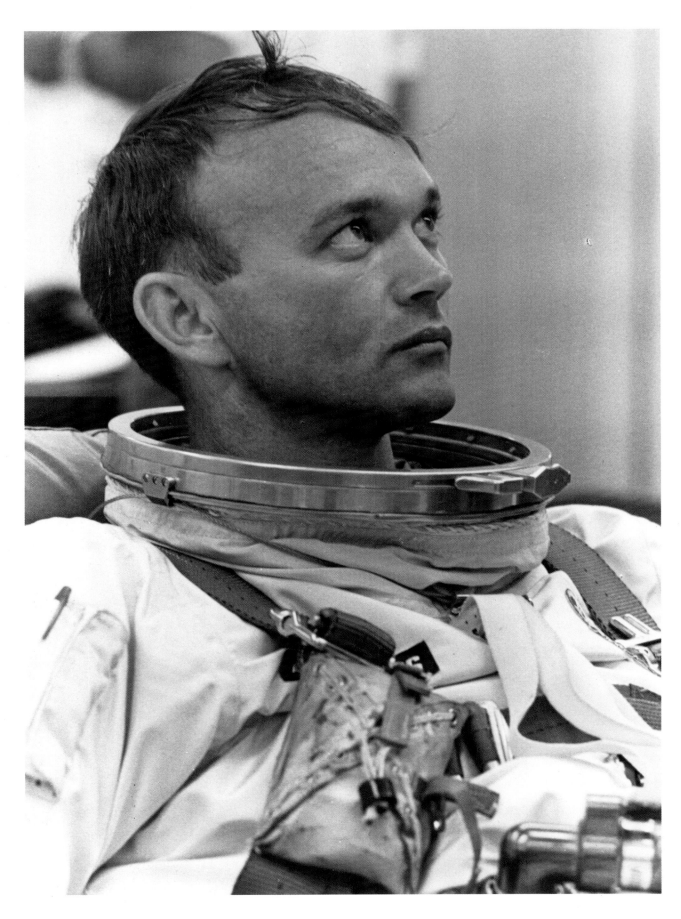

after forty-five orbits, and *Gemini 9* landed in the Atlantic Ocean less than two miles (3.2 kilometers) from the predicted point of splashdown.

Six weeks later, on July 18, both the *Gemini 10* and its *Agena* target vehicle were orbited after flawless launches by each carrier from Cape Kennedy. The 8,248-pound (3,741 kilogram) *Gemini 10* and its two astronauts, Commander John W. Young and Major Michael Col-

lins, first entered into a relatively low orbit with a perigee (the point in an orbit where the Moon or spacecraft is closest to the Earth) of ninety-nine miles (160 kilometers). Preparations were made to raise the perigee and attempt a rendezvous with the *Agena 10* target. This was achieved after over six hours and the expenditure of an excessive quantity of propellant. Docking was successful, however, and using *Agena*

10's propulsion system, *Gemini 10*'s perigee was raised to 184 miles (297 kilometers), and its apogee (the furthest or highest point away from the Earth) to over 476 miles (767 kilometers). Later, rendezvous was also made with the dormant *Agena* target vehicle left over from the *Gemini 8* mission.

During his successful thirty-eight minute EVA, Collins remarked, "My God, the stars are everywhere." He enjoyed his time, using a hand-held maneuvering unit similar to the one used in the *Gemini 4* mission. He performed a number of tasks on his spacecraft and then went over to the *Agena 8* target to remove some micro-film and to replace an instrument package for detecting micro-meteoroids. This exchange was another important first that signalled a new technique in space research.

The *Gemini 10* mission splashed down after a flight of seventy hours and forty-six minutes. The capsule landed only five miles (eight kilometers) from the air carrier *Guadalcanal* in the Atlantic Ocean.

Gemini 11 was launched with great precision on September 12, 1966, less than a second after programmed and ninety-seven minutes after the *Atlas-Agena* launched the *Agena* target vehicle. The 8,509-pound (3,860-kilogram) *Gemini 11* capsule carried Commander Charles Conrad Jr. and Lieutenant Commander Richard Gordon into an orbit that ranged from 100 miles (161.9 kilometers) to 175 miles (282 kilometers) in altitude. Gordon's EVA went smoothly, although like Cernan in *Gemini 9*, Gordon's over-exertion outside the craft resulted in an overload of the life support system. During the course of the flight, the two astronauts observed Russia's *Proton 3* satellite, which had been launched on July 6.

Gemini 11 returned to Earth in a completely automatic reentry maneuver, established by the IBM on board computer. Splashdown occurred less than two miles (3.2 kilometers) from the projected target point on September 15. Recovery of this craft was made by the U.S.S. *Guam*.

The twelfth and final spacecraft in the *Gemini* series carried Captain James A. Lovell (it was his second *Gemini* flight) and Major Edwin E. Aldrin. Both *Gemini 12* and its *Agena* target vehicle were launched from Cape Kennedy on November 11, 1966; the *Gemini 12* splashed down in the Atlantic Ocean on the 15th of November, ninety-four hours and thirty-four-and-a-half minutes after lift-off.

All of the planned experiments were conducted, although some equipment malfunctions and other difficulties occurred: Four of the thrusters used for maneuvering in space failed to operate. Rendezvous and docking with the *Agena* were performed several times. The highlight of the flight was a space

walk by Aldrin, building up the total EVA time by American astronauts to about twelve hours. Aldrin performed a number of tasks, including cutting wire, making lock connections, hooking and unhooking spring-loaded snap hooks, and using a torque wrench. In order to maintain his body position, he used portable Velcro handholds and foot restraints, plus tethers leading from his spacesuit, during the EVA.

While the space walks and the docking of the two vehicles were perhaps the most dramatic features of the *Gemini* program, each flight conducted many different experiments, including studying the effect of prolonged weightlessness on humans, the development of rendezvous and docking techniques, the ability of the astronaut to leave his spacecraft and landing techniques. The *Gemini* astronauts took hundreds of vivid color photographs of the Earth to assist geologists, oceanographers, geophysicists and meteorologists. They tested their ability to spot patterns on the ground from space. They conducted a series of medical experiments and measurements to determine whether the heart and bones are weakened by long space flights, and whether mental capability changes in the stress of space. Radiation and charged particles were measured, and an attempt was made to communicate with the ground using a laser. These and other experiments gave a generally optimistic picture of man's ability to survive and function in space. As far as could be determined, flights of up to two weeks did not damage the human body or man's ability to make observations and evaluations.

As the last *Gemini* returned to Earth, the *Agena* crafts were left orbiting until they decayed and the craft plunged, burning up as it raced through the atmosphere.

NASA's next step was the *Apollo* missions. By now, the United States was learning what an intense joy a comprehensive space program could be.

Gemini 12 command pilot James Lovell carries an over-sized flight ticket signed by launch crew personnel as he and pilot Edwin Aldrin (behind him) walk to the elevator at Cape Kennedy. Launched on November 11, 1966, this was the last mission in the Gemini Program.

Apollo astronaut Edgar D. Mitchell, the Lunar Module (LM) pilot. The *Apollo 14/Saturn V* space vehicle carrying astronauts Alan B. Shepard Jr., Stuart A Roosa and Edgar D. Mitchell was NASA's sixth manned expedition to the Moon. Astronauts Shepard and Mitchell descended in a Lunar Module to the Moon's hilly upland region north of the Fra Mauro crater. During the thirty-three hours Shepard and Mitchell spent on the lunar surface, Roosa piloted the Command Module in lunar orbit. Close to ten of those hours were spent outside the LM.

CHAPTER 3

Oh, I have slipped the surly bonds of Earth
And danced the skies on laughter-silvered wings;
Sunward I've climbed, and joined the tumbling mirth
of sun-split clouds—and done a hundred things
you have not dreamed of—wheeled & soared & swung
High in the sunlit silence. Hov'ring there,
I've chased the shouting wind along, and flung
My eager craft through footless halls of air
Up, up the long, delirious, burning blue
I've topped the wind swept heights with easy grace
Where never lark, or even eagle flew
And, while with silent, lifting mind I've trod
The high untrespassed sanctity of space,
Put out my hand, and touched the face of God.

—JOHN GILLESPIE MAGEE JR.

"Houston, Apollo 11…I've got the world in my window."

—ASTRONAUT MICHAEL COLLINS

TO DREAM THE IMPOSSIBLE DREAM

The Apollo program was based on the development of the *Saturn* carriers. Large as they were, *Saturns 1* and *1B* were only preludes to the more powerful carrier that was needed to fulfill the goal set by President Kennedy—landing an American on the Moon within the decade—in his speech to Congress on May 25, 1961: "With the advice of the Vice President…we have examined where we are strong and where we are not, where we may succeed and where we may not…Now is the time to take longer strides—time for a great new American enterprise—time for this nation to take a clearly leading role in space achieve-ment, which in many ways may hold the key

to our future on Earth."

In order to accomplish this dream, a NASA Defense Department Executive Committee for Joint Lunar Study and a Joint Lunar Study Office were established. Some very basic decisions had to be made before a spacecraft could even be designed. Problems such as how many crew members were needed for the trip, how much equipment they would need in order to survive and to perform scientific experiments and other pertainent questions needed to be answered.

As mentioned before, NASA considered three flight plans for a Moon landing and after much consideration chose the lunar orbital rendezvous (LOR) during July of 1962; thus work on a final design for *Apollo* began.

Nevertheless, even as decisions regarding *Apollo* were being made, accelerated work continued on the *Saturn* launch rocket to develop what would become the *Saturn 5*. This work was accomplished at the Marshall Flight Center in Huntsville, Alabama. On January 25, 1962, NASA approved a development program for the carrier, which was given the highest priority. *Saturn 5* was to have three stages: the S-1C stage, the S-2 stage and the already existing S-4B stage from the *Saturn 1B*.

The S-1C stage developed by the staff at the Marshall Center, with the support of the Boeing Company, was tuned for production assembly at the large NASA-owned Michaud plant in New Orleans, Louisiana. The S-1C stage was about 128 feet (thirty-nine meters) tall and thirty-three feet (ten meters) in diameter. The 1969 model used for *Apollo* weighed nearly 300,000 pounds (136,079 kilograms) empty and held 4.7 million pounds (2.1 million kilograms) of liquid oxygen and RP-1 kerosene fuel. *Saturn 5*'s first stage thrust was about 7.7 million pounds (3.5 million kilograms), with each of its F-1 engines capable of developing more than the 1.5 million pounds (680 thousand kilograms) of thrust generated by the *Saturn 1B*.

The third stage was the S-4B, 58.1 feet (17.7 meters) long and 21.7 feet (6.6 meters) in diameter and powered by a J-2 engine whose thrust variance during flight could range from 184,000 to 230,000 pounds (83,461 to 104,327 kilograms). When put together, the *Saturn*

Opposite page: Moving along a specially constructed crawlway capable of supporting loads of approximately eighteen million pounds (eight million kilograms), the Transporter, a mobile launcher, carries the *Apollo 4* towards its launch site. *Apollo 4* stood 364 feet (112 meters) high and weighed 6,286,000 pounds (208,000 kilograms) when fueled. The Transporter was used to move the craft from the Vehicular Assembly Building to its launch site. *Left:* This aerial photograph taken in 1977 illustrates many of the facilities at Launch Complex 39, the prime launch and recovery site for the Space Shuttle. Dominating the view is the Vehicle Assembly Building, which towers 525 feet (162 meters) high. The low structure to the right is the Launch Control Center.

Right: The Apollo spacecraft is hoisted to its test chamber in the manned Spacecraft Operations Building at Kennedy Space Center. *Opposite page:* May 25, 1961. President Kennedy delivers his historic message to a joint session of Congress. Kennedy passionately announced, "I believe this nation should commit itself to achieving the goal, before this decade is out, of landing a man on the Moon and returning home safely to Earth." This goal was achieved by astronaut Neil A. Armstrong on July 20, 1969 at 10:56 p.m. (EST), when he became the first man to set foot on the Moon.

5 stood 363 feet (110.6 meters) tall. Fully fueled it weighed nearly 6.4 million pounds (2.9 million kilograms). It was able to send a spacecraft to the Moon, or place a 150-ton (136-metric ton) payload into orbit around the Earth.

In order to accommodate the gigantic carrier, an equally immense facility had to be built at Cape Kennedy. The 54-story, 526-foot high (160.3 meters) vertical assembly building had 130 million cubic feet (3,679,000 cubic meters) of space, making it the world's largest building. It contained four huge bays, where four carriers and their spacecraft payloads could be erected and assembled.

The more than $20 billion that the *Apollo* project would cost aroused considerable opposition, and NASA had to garner public acceptance for the Apollo Program. Thus, NASA Deputy Administrator Hugh L. Dryden pointed out in June 1961 that the money for *Apollo* "would not be spent on the Moon ... [but rather] in the factories, workshops and laboratories of our people for salaries, new materials and supplies which, in turn, represent income for others." Vigorous support for the program came from President Kennedy, who, in his 1962 State of the Union Address, said that America's aim was "not simply to be first on the Moon ... Our objective in making this effort, which we hope will place one of our citizens on the Moon, is to develop a new frontier in science, commerce and cooperation, the position of the United States and the free world. This nation belongs among the first to explore it. And among the first, if not the first, we shall be."

North American Aviation Inc. (now North American Rockwell Inc.) was awarded a contract for the *Apollo* reentry capsule and its attendant deep space propulsion unit, consisting of the command and service module (CSM), before the LOR approach was selected. In November 1962, Grumman Aircraft Engineering Corporation was given a contract for the Lunar Excursion Module (LEM), in which the two astronauts were to touch down on the Moon. This module later came to be known as the Lunar Module (LM). North American continued to be responsible for the other two modules.

The three men in the *Apollo* crew

spent most of the period required for a lunar mission (upwards from eight days) in the Command Module (CM), the only part of the *Apollo* spacecraft that returns to Earth. Nearly eleven feet (3.4 meters) high and thirteen feet (3.96 meters) in diameter at its base, the conical module weighed approximately 12,000 pounds (5,443 kilograms) with the crew aboard. The CM served as the flight control center, living quarters and reentry vehicle at the end of the mission. It is divided into three compartments: forward, crew and aft.

The forward compartment was built around the tunnel that connects the command and Lunar Modules when they are docked. It contained para-

chutes, recovery antennas, the beacon light, the "sling" for retrieving the capsule after it lands in the sea, reaction control engines and a mechanism for jettisoning the forward heat shield during reentry so that the parachutes can be deployed.

The crew compartment allotted the astronauts nearly 210 cubic feet (5.94 cubic meters) of living space, vehicular controls and display panels along with other operational equipment. It was fitted with two hatches, the first employed for normal entering and exiting and the second for moving into and, later, back out of the Lunar Module once the *Apollo* had entered into orbit around the Moon. In addition to these hatches,

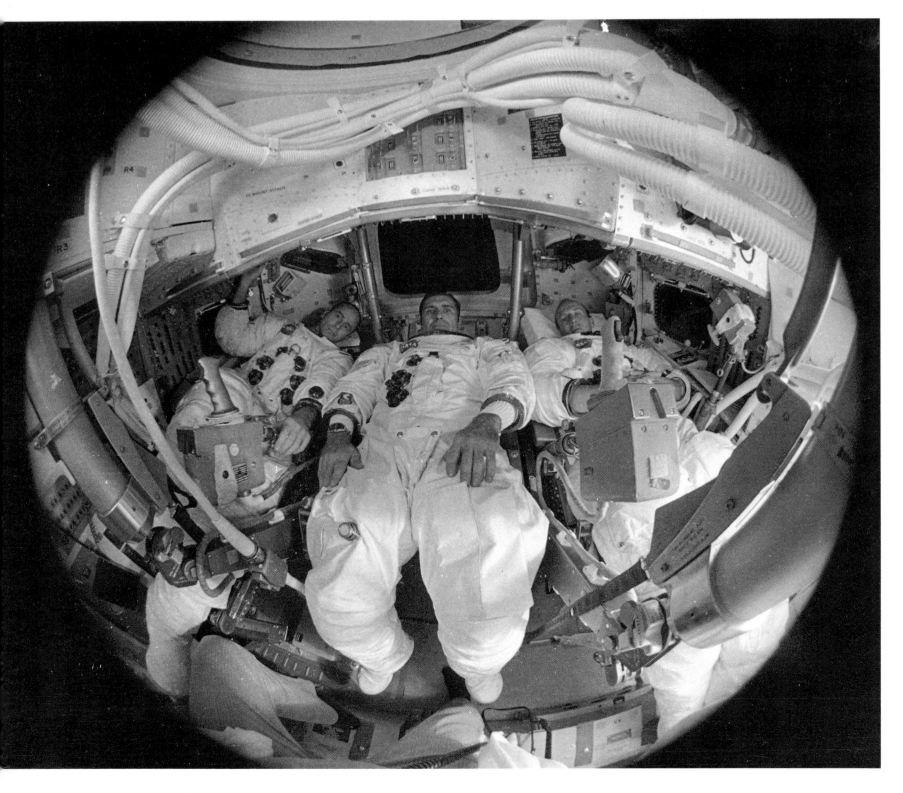

Opposite page: Apollo 12 astronauts look on while their spacecraft is checked out at the Kennedy Space Center Operations Building. They are (from left to right): Alan L. Bean, Lunar Module pilot; Richard F. Gordon, Command Module Pilot; and Charles Conrad Jr., commander. *Left:* This photograph of the Moon was taken when the *Apollo 10* was high above the lunar equator, near twenty-seven degrees east longitude. Apollo landing site 3 is on the lighted side, in a dark area just north of the equator: Apollo landing site 2 is near the lower left margin of the Sea of Tranquility—the large dark area near the center of the photograph.

The first stage of the *Saturn V/Apollo* 6 mission falls away from the second stage after 148 seconds of burn time. *Apollo* 6 was the second unmanned Apollo flight, conducted to ensure that the rockets would be safe for manned flights to the Moon.

the crew compartment had two side windows, two front windows used in rendezvous and docking operations and a hatch window. Each window had both inner and outer panes—the former of tempered silica glass, the latter of amorphous-fused silicon. All ultraviolet and most infrared radiation was filtered out by these windows.

Inside the crew compartment, the astronauts spent most of their time on form-contoured padded couches, but two people could stand if the seat portion of the center couch was stowed. Two astronauts at a time could sleep in sleeping bags located underneath the left and right couches. Food and water, as well as waste elimination devices, were kept in storage bays around the walls of the compartment. The environmental control system maintained a constant temperature of seventy-five degrees F. (twenty-four degrees C.) and an atmospheric pressure of five pounds per square inch. Normally, the crew did not wear spacesuits, donning them only during the launch, docking, crew transfer and reentry phases of the mission.

Although the three astronauts were cross-trained, the flight commander (who occupied the left couch) generally operated the flight controls; occasionally the Command Module pilot (in the center couch) would take over.

The latter's principal jobs were to guide and navigate the spacecraft along its trajectory and, once the *Apollo* had been placed in orbit around the Moon and the Lunar Module detached, to monitor the lunar surface. On the right-hand couch was the Lunar Module pilot; while in the Command Module he managed all subsystems. In the event of an emergency, the Command Module could be handled by any of the three astronauts.

The final compartment of the Command Module, located aft above the rear heat shield, was divided into twenty-four bays containing propellant, helium and water tanks, ten reaction control engines, the wiring and umbilical connections with the service module, part of the impact attenuation system and various other necessary instruments.

The inner shell of the Command Module was an aluminum sandwich structure with a welded inner skin, a bonded honeycomb core and an outer sheet. The outer heat shield was a

three-piece structure of brazed honey-comb stainless steel bonded to a phenolic epoxy resin (derived from coal tar) that absorbs heat by burning away, or ablating, as the module descends through the Earth's atmosphere after the lengthy lunar voyage. The shield weighs about 3,000 pounds (1,335 kilograms) and varies in thickness from one-half inch (1.25 centimeters) to two inches (five centimeters).

Mounted directly behind the Command Module was the service module (SM), a cylindrical unit more than twenty-four feet (7.3 meters) long and nearly thirteen feet (four meters) in diameter. It weighed 11,500 pounds (4,123 kilograms) when completely empty and close to 55,000 pounds

(24,948 kilograms) when loaded with propellant. This module contained the principal propulsion system of the *Apollo* spacecraft, propellant, electrical system, water and other supplies.

The CM and SM fly together as a unit until the end of the lunar round trip, when the SM is jettisoned prior to atmospheric entry. The propulsion system of the SM operates on nitrogen tetroxide and a hypergolic (self-igniting) mixture of hydrazine and unsymmetrical dimethylhydrazine. The engine developed 20,500 pounds (9,299 kilograms) of thrust and was used to effect all major changes in velocity once the craft has left its temporary "parking" orbit around the earth and begun its voyage to the Moon. Developed by the Aerojet General Corporation, it could

be restarted up to fifty times during a mission. Although the engine had no throttle, it could provide small bursts of power by burning for as little as 0.4 seconds at one time. Under normal conditions the engine received commands for thrusting from a computer located in the CM, but, if the need arose, it could be operated manually by the astronauts.

The module in which two of the three astronauts descended onto the Moon's surface was the Lunar Module (LM). During flight out to the parking orbit, the module is housed in the spacecraft inside the Lunar Module Adapter (SLA), a twenty-eight foot (8.53-meter) tapered cylinder between the SM and the instrument unit located above the third stage of the *Saturn 5* carrier.

Opposite page: A suborbital test of the Apollo capsule escape system. *Below* (left to right): Edward H. White, senior pilot, Roger B. Chaffee, pilot and Virgil I. Grissom, command pilot practice for the first manned Apollo mission in the Apollo Mission Simulator. Visible in the upper left of the picture is the four and one-half television camera which was used on the flight. Utilizing integrated circuitry, the TV camera relayed "live" images of the mission back to Earth.

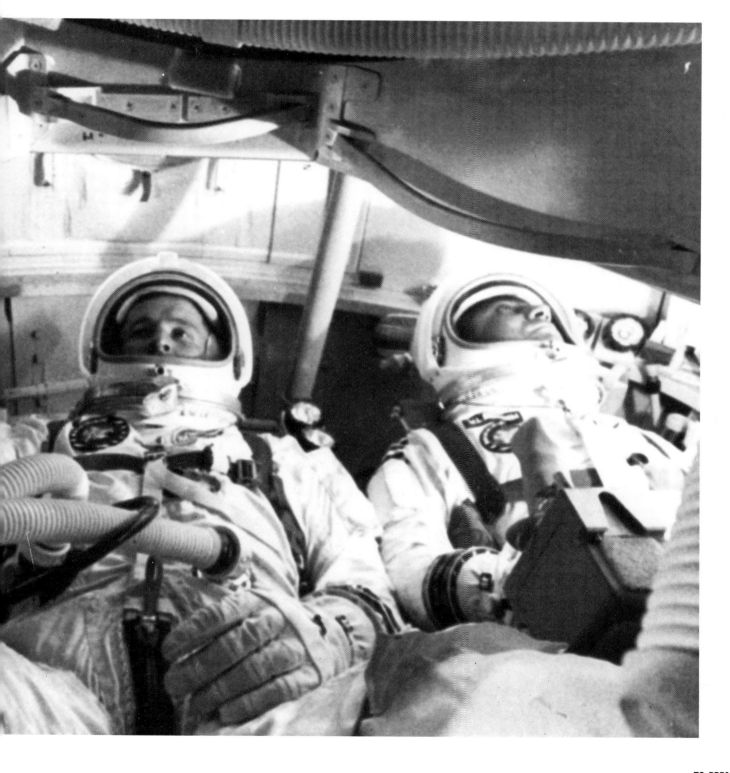

Sixteen NASA astronauts participate in three days of tropic survival training at Albrook Air Force Base in the Panama Canal Zone. *Inset:* Displaying the outfits they fabricated for desert survival training exercises at Stead Air Force Base in Nevada are (left to right): Astronauts Neil A. Armstrong, Frank Borman, Charles Conrad Jr., James A. Lovell Jr., James A. McDivitt, Elliot M. See Jr., Thomas P. Scafford, Edward H. White II, John W. Young and Donald K. Slayton.

Right: The United States naval ship *Mercury* was one of three "ground" support ships which helped the *Apollo* craft reach the Moon. This ship was equipped with four dish antennas as well as several helix and whip antennas to monitor vital craft information. *Below:* One of three "firing rooms" at the Launch Control Center (LCC). The three rooms have some 450 consoles which contain controls and displays required for the checkout process and fifteen display systems in each firing room. Each system is capable of providing digital information instantaneously. Sixty TV cameras, transmitting pictures on ten modulated channels, are positioned around the *Apollo.* LCC also contains 112 operational intercommunication channels used by the crew in the checkout and launch countdown.

After leaving the parking orbit, the combined command and service modules separate from the combined LM, SLA and S-4B and turned around to dock with the LM so that the astronauts can later transfer to the landing vehicle.

Once docking is completed, the LM is pulled out of the SLA, whose panels are then jettisoned, and the CSM and LM continued onward toward the Moon. The S-4B, meanwhile, is sent into a trajectory that puts it into an orbit around the sun, where it will soon burn up.

Built by the Grumman Aircraft Engineering Corporation, the Lunar Module had two distinct parts: the descent stage and the ascent stage. Using the rocket engine of the descent stage to brake its landing, the LM would settle softly into the lunar surface, where it served as the astronauts' temporary home and base of operations, their communications center and their supply compartment. The descent stage became a stationary launch platform from which the ascent stage, with the two astronauts on board, took off for lunar orbit and subsequent rendezvous and docking with the waiting *Apollo* CSM (Command and Service Module combination).

A number of lunar module vehicles were constructed as part of the R&D program. First came the M series,

which consisted of mock-ups; then the TM, or test series; the LTA, or test article series; and, finally, the LM "man-rated" series, which was deemed safe for astronauts.

Early LM vehicles were tested in Earth Orbit, where Moon flight conditions were simulated. The total pressurized volume of the LM was 235 cubic feet (five cubic meters), of which 160 cubic feet (four cubic meters) was habitable space.

Flight testing for the *Apollo* manned mission to the Moon began in November 1963, when the launch escape system of a so-called boilerplate module, BP-6, was tested in the desert at White Sands, New Mexico.

TRAGEDY STRIKES

In 1964 and 1965, flights were made with Little Joe solid-propellant rockets. Also, during the *Saturn 1* program, engineering test modules of the spacecraft were sent into orbit. By the end of February 1966, *Apollo* had progressed to the more powerful *1B*, which carried a production model of both the command and service modules along a suborbital flight pattern. Some 600 measurements (including thrust, fuel use and communications) radioed back to the ground satisfied NASA engineers that things were going quite well. Other *1B* tests in July and August were also satisfactory.

Then tragedy struck. It didn't strike in the air or in the far reaches of space; instead, tragedy struck NASA on the ground. The nation, overconfident from repeated successes, was, to say the least, crestfallen.

On January 27, 1967, at Cape Kennedy, astronauts Virgil I. Grissom, Edward H. White and Roger B. Chaffee died in a flash fire that destroyed an *AS-204* spacecraft undergoing routine tests at Launch complex 34. The three were to have been the first astronauts in the *Apollo* manned flight program.

As the nation grieved, research and testing at NASA was almost completely halted while an appointed committee studied the available facts concerning the accident and filed a report. The report concluded that the space quest must continue and that NASA should resume its attempt to meet its goal of putting an American on the Moon.

Arlington Cemetery, Virginia. The flag-draped coffin of astronaut Virgil I. Grissom is escorted by fellow astronauts (left) Donald K. Slayton, Alan B. Shepard Jr., (right front) John Glenn Jr., Leroy G. Cooper and John W. Young. *Inset:* Leaving the Apollo Mission Simulator are (bottom to top) Virgil Grissom, Edward White and Roger Chaffee. These three astronauts were to have been the first manned-mission pilots, until a flash fire aboard an AS-204 spacecraft killed them.

The *Apollo* 9 LM seen from the Command Service Module. On this fifth day of the *Apollo* 9 Earth orbital mission, the craft is in a lunar-landing configuration. The landing gear of the spider has been deployed and Lunar surface probes (sensors) extend out from the landing gear foot pads. Inside this craft were astronauts James A. McDivitt, commander, and Russell L. Schweickart, lunar module pilot. Astronaut David R. Scott, command module pilot, remained at the controls in the CM "Gumdrop."

Testing immediately resumed, and on October 11, 1968, another NASA first occurred—*Apollo 7* was launched, with Captain Walter Schirra Jr., Major Donn Eisele and Walter Cunningham on board. This flight, which completed 163 Earth orbits, was the first launch by NASA using an upper-stage rocket that had not been originally designed to carry military warheads. Peaceful use of space had indeed come a long way, despite the fact that the other stages of the *Saturn 1B* had emerged from the military IRBM and ICBM program.

Apollo 7 was destined to be the only manned spacecraft launched by a *Saturn 1B*, for *Apollo 8* and all subsequent *Apollo* missions would require a launch vehicle capable of hurling the craft all the way to the Moon. This performance required the *Saturn 5*, which ultimately boosted all of the American astronauts toward their lunar missions.

As 1968 drew to a close, NASA launched *Apollo 8*, which orbited the Moon ten times with Colonel Frank Borman, Captain James Lovell Jr. and Lieutenant Colonel William Anders on board. The mission went extremely smoothly.

On February 24, 1969, NASA launched the *Mariner 6*, which completed its exploratory fly-by of Mars on July 31. In March, NASA successfully launched *Apollo 9*, the first flight of the complete spacecraft, and from May 18 through May 26, NASA launched what was referred to as "the full dress rehearsal" for the awaited Moon landing. On board this *Apollo 10* flight were Colonel Thomas P. Stafford, Commander John W. Young and Commander Eugene Cernan. This dress rehearsal went extremely well, and all was set for the final step—landing and walking on the Moon.

On July 16, 1969 *Apollo 11* lifted off with astronauts Neil A. Armstrong, Lieutenant Colonel Michael Collins and Colonel Edwin E. Aldrin Jr. for the mission that a generation earlier had seemed like science fiction. Four days later, the Lunar Module separated and descended to the lunar surface. The descent, which was manually controlled, had fuel problems, and with only a few seconds of precious fuel remaining, Neil Armstrong announced, "Houston . . . Tranquility Base here . . . The Eagle has landed."

Opposite page: Astronaut Walter Cunningham's two children, Brian, 8, and Kimberly, 5, watch the activity inside the Mission Operations Control Room during a visit to the Mission Control Center. *Left:* In this photograph taken by Neil Armstrong, Edwin E. Aldrin Jr., the lunar module pilot, is seen walking near the lunar module during the *Apollo 11* EVA. A LM foot pad can be seen in the lower right.

Pictured here with a container of lunar soil collected during the *Apollo 12* EVA is an astronaut wearing a checklist on his left wrist to facilitate following a pre-planned pattern. The astronaut who snapped the photo is reflected in the face shield of his fellow crewman.

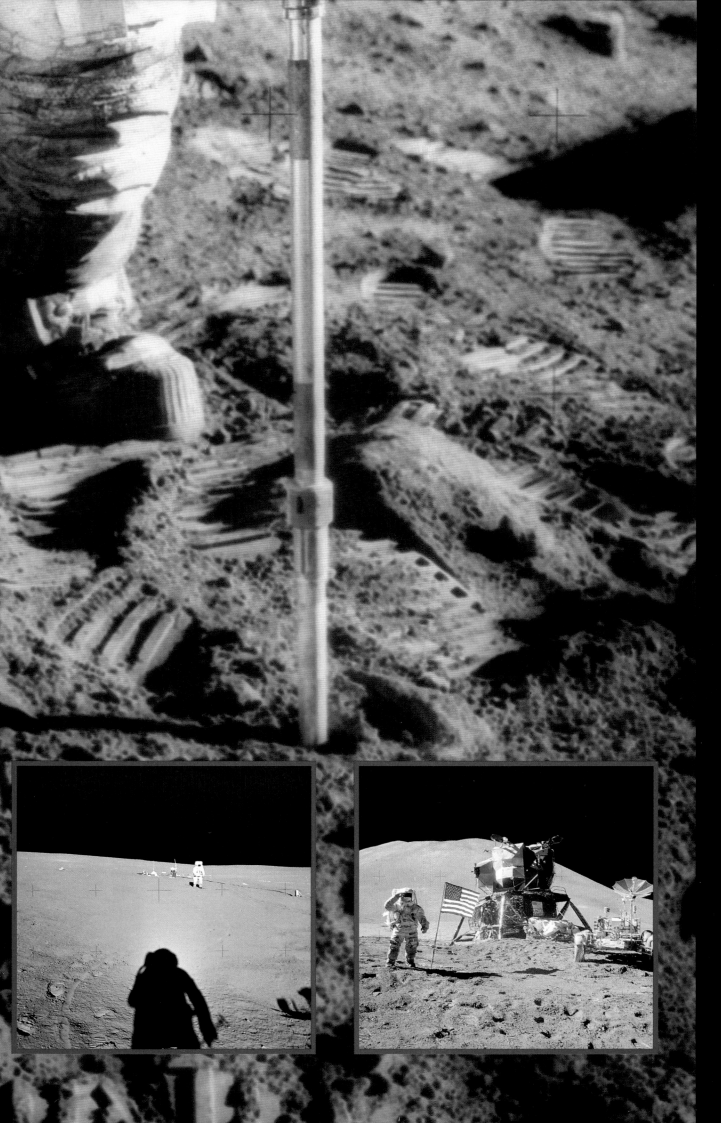

Apollo 12 astronaut Alan L. Bean taking a core sample during an EVA on the lunar surface. Footprints made by the two *Apollo 12* crewmen are seen in the foreground. *Apollo 12* was the second manned lunar landing mission.

Inset, left to right:

The LM is heading for its landing site, the Ocean of Storms. The large crater on the right is named Herschel.

Apollo 15 astronaut James Irwin, lunar module pilot, next to the Rover, looking northeast with Mount Hadley in the background. The mission stayed at the Hadley-Apennine site a total of sixty-six hours and fifty-five minutes.

Apollo 14 astronaut Edgar Mitchell, the LM pilot, strolls toward the LM from the deployment site of the Apollo Lunar Surface Experiments package. The shadow of Alan Shepard, Jr., who took the photo is seen in the foreground.

Apollo 15 Astronaut Irwin saluting the flag. On the right is the lunar Rover; Hadley Delta is in the background.

A beautiful view of the desolate lunarscape. Scientist-astronaut Harrison H. Schmitt is shown working at the Lunar Roving Vehicle during the second *Apollo* 17 EVA at the Taurus-Littrow landing site. This is the area where Schmitt first spotted the orange soil, clearly visible on either side of the Rover.

In progress is the largest ticker-tape parade in the history of New York City. This parade was for national heroes Neil Armstrong, Michael Collins and Edwin Aldrin after their return from the historic *Apollo 11* flight.

Left behind on the lunar surface is a family photograph showing astronaut Charles M. Duke Jr., his wife and two children. Duke was the *Apollo 16* lunar module pilot.

Hours later, millions of television viewers watched Armstrong step onto the lunar surface and make his now famous speech: "One small step for man; one giant step for mankind."

As Edwin E. Aldrin Jr. joined Armstrong on the gray powdered surface, collecting samples of Moon rocks, and Michael Collins circled in the *Columbia* Command Module, a nation rejoiced—the ten-year national commitment had been met.

Other *Apollo* flights also met with resounding success. *Apollo 12* discovered a Moon site, the Ocean of Storms, where an unmanned United States craft, the *Surveyor 3*, had been sitting for almost three years. Astronauts Charles Conrad and Alan L. Bean removed several pieces, including a television camera that NASA scientists would analyze to learn what thirty months of lunar exposure had uncovered.

The *Apollo 15* mission included a Lunar Rover. This electric-powered, four-wheel-drive vehicle enabled astronauts David R. Scott and James B. Irwin to explore more of the Moon's landscape than was previously possible. The seventh and final mission, *Apollo 17*, was an extremely productive scientific venture, exploring the Taurus-Littrow site, where both the oldest and the newest moon rocks were found.

When the *Apollo* project officially ended, more than $123.5 billion had been spent, and NASA had put twelve men on the Moon within less than twelve years.

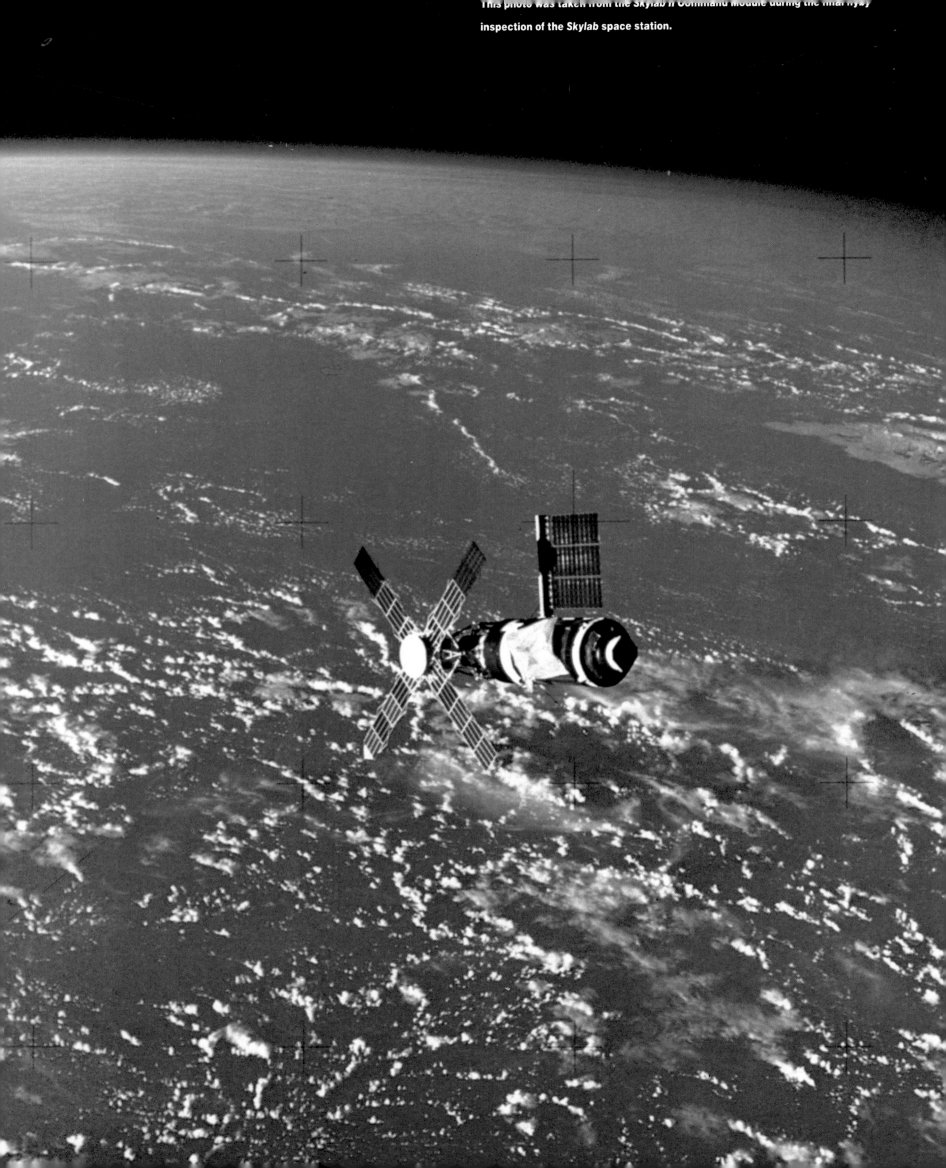

CHAPTER 4

"Unlimited vacuum of outer space and the absence of gravitational forces in an orbiting satellite, make an Earth-circling spacecraft an ideal site for the observation of stars, the Moon, the Sun, the planets and, particularly, our Earth."

—PROFESSOR HERMANN OBERTH

SKYLAB: A WORKSHOP IN SPACE

While the United States was busy landing men on the Moon, the Soviet Union continued to launch its spacecrafts under its Kosmos umbrella organization (which covered military as well as weather and observation satellites). This led to the June 1971 launching of the Soviets' *Soyuz 11* spacecraft mission, which docked with their *Salyut 1,* an embryonic space station. When cosmonauts Vladislav N. Volkov and Viktor I. Patsayev crawled through the airlocks into *Salyut 1* (Commander Georgi Dobrovolski stayed aboard the *Soyuz*) it became the first manned laboratory in history. Beginning June 7, for a period of twenty-three days these cosmonauts carried out their respective duties, orienting themselves to long periods

Right: At a training session in the *Soyuz-T* spaceship are (left to right): French cosmonaut Jean-Loup Chretien, USSR pilot-cosmonaut Vladimir Dzhanibekov and USSR pilot-cosmonaut Alexander Ivanchenkov. *Opposite page, left to right:* The crew of the first manned *Skylab* mission, Joseph P. Kerwin, science pilot; Charles Conrad Jr., commander; and Paul J. Weitz, pilot. They are pictured with a model depicting the *Skylab* space station cluster.

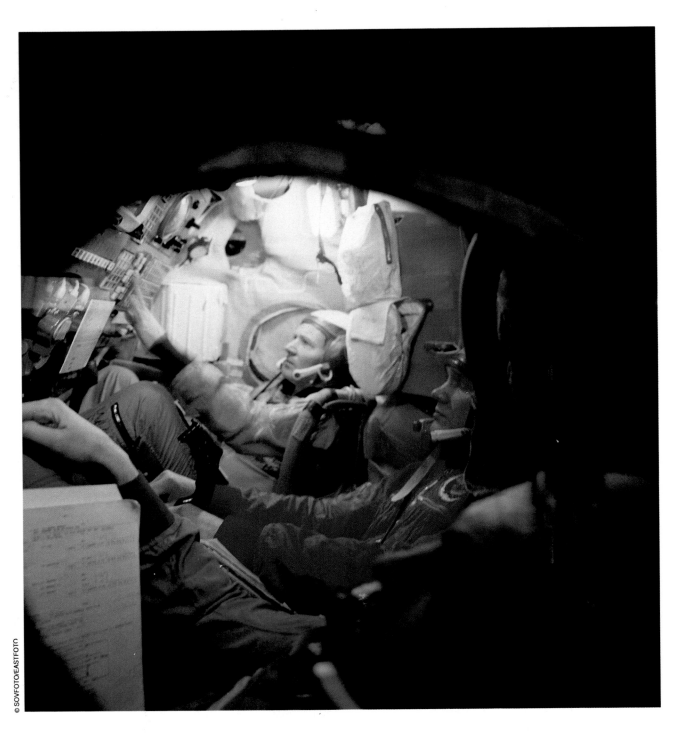

© SOVFOTO/EASTFOTO

in space, learning how to maneuver and navigate, investigating atmospheric conditions and Earth resources and conducting research in space physics, space biology and space medicine.

On June 29 the proud cosmonauts left *Salyut 1* and reentered *Soyuz 11,* rejoining their commander. The next morning they undocked from the *Salyut 1* and initiated their reentry sequence. At the moment the spacecraft's explosive bolts were fired (which caused the separation to commence), all communications went dead. After what appeared to be a routine, automatically controlled reentry, the landing capsule touched down on Russian soil. An already nervous helicopter recovery team (communication lines were still dead) sped to the capsule and opened it, finding, to their horror and

the world's, that all three cosmonauts had returned to Earth dead.

As the world grieved, the Russians traced the cause to the accidentally opened exhaust valve, probably triggered by the explosive bolts during the separation phase. Apparently, there had not been adequate time for the crew to attempt to close the valve before the capsule's air escaped. George Low, NASA's deputy administrator, called the accident "a terrible tragedy," adding that the three cosmonauts "were pioneers in their achievements in space in establishing the first manned space station."

As a result of this tragic event, the Soviet Union took a long respite from its man-in-space program. Eventually, the Soviets attempted to launch *Soyuz 12* on April 8, 1973; inexplicably, it broke up into hundreds of pieces once in space.

The accident's cause was never revealed, and the official Soviet announcement proclaimed only that "the fulfillment of the flight program of the *Salyut 2* orbiting station is complete." In 1974, a much redesigned *Salyut 3* was orbited and successfully carried out biological, medical and space research and geological and engineering experiments.

AMERICA'S LABORATORY IN SPACE

Meanwhile, the United States, having completed the highly successful Apollo lunar exploration program, resumed manned Earth orbital activities with its *Skylab*, NASA's early space station design.

The origin of *Skylab* can actually be traced back to November 1962, when the Douglas Aircraft Company (now known as the McDonnell Douglas Corporation) published a briefing manual entitled "1965 Manned Space Laboratory." In this manual, engineers called for modifying the *Saturn 1B* rocket's second (S-4B) stage into an orbiting laboratory. After the stage had completed its propulsion role, an astronaut work crew, orbited by another *Saturn 1B*, was to go on board and convert the now empty liquid hydrogen tank into a habitable laboratory by pumping in the necessary breathable oxygen. From this, such terms as "spent stage" and "wet workshop" evolved (see page 101), the latter derived from the fact that

This photo taken by the *Skylab* 2 Command Service Module (CSM) is of the *Skylab* space station cluster in Earth orbit. The area below the cluster is clouds over water. Note the deployed parasol solar shield which shades the Orbital Workshop—the area of the missing micrometeoroid shield. The other OWS solar system wing has been successfully deployed. The OWS solar panel on the opposite side is completely missing.

residual, unused liquid hydrogen was to be drained from the fuel tank before it could accommodate the crew. NASA engineers at the Marshall Space Flight Center had also been thinking of developing a manned, orbiting laboratory and encouraged the Douglas study.

From the beginning, the idea behind *Skylab* was to utilize the hardware and techniques that emerged from the Apollo experience. An early Apollo Extension System Program was established to synthesize and coordinate all Apollo-related concepts and proposals beyond the initial series of lunar landing missions. This was followed in August 1965 by the Apollo Applications Program (AAP), set up within the Office of Manned Space Flight. On December 1 of the same year, NASA gave the formal go-ahead for *Skylab* (then referred to as Orbital Workshop Skylab, or OWS) as a major element of the AAP. Scheduling called for OWS development to proceed concurrently with that of Apollo itself, with the Marshall Center team taking prime responsibility for hardware development and the Manned Spacecraft Center group coordinating all activities involving the *Apollo* CSM (Command Service Module) ferry craft and the astronauts (see page 76). In 1967, the AAP's director, Charles W. Matthews, summed up the philosophy behind the program:

"The activities involved in [AAP] represent major steps in the utilization of our space exploration and applications. In particular, increased knowledge on the effective integration of men into the total system should accomplish much in determining the character, systems configurations and operational approach in future programs. The ability to capitalize on the large investments already made in the Apollo program affords the opportunity to carry on this [AAP] work in an efficient and economical manner."

NASA asked the world's leading scientists to identify the most promising objectives for the proposed orbital workshop. For instance, solar physicists were approached by NASA for their input as to how the OWS should be equipped to answer their questions about the Sun and its activities, and the aerospace medical community was queried as to what experiments could be conducted aboard the station to help answer their questions about

The Soviet spacecraft, *Soyuz*, photographed from the *Apollo* spacecraft. These two crafts made a historic rendezvous in space on July 17 and 18, 1975.

man's ability to live and perform in zero gravity.

As anticipated, the response to this solicitation was overwhelming. Several orbital workshops could easily have been filled with all the proposed experiments (in fact, four Skylab missions were eventually carried out). NASA, supported by the National Academy of Sciences, screened the proposals, not only from the point of view of scientific desirability, but also with regard to the demands on the station's limited facilities and the time of its crew. The "time line" factor was very important, and it was this more than anything else that weighed heavily against the concept of a "wet workshop," which would have required the astronauts to spend a tremendous amount of their valuable orbital time simply moving their living and lab equipment into what, during the launch phase, would have served as a liquid ("wet") hydrogen tank.

Budget cuts and program stretch-outs between early 1967 and mid 1968 led to NASA's decision that the OWS would have to follow the Apollo lunar landing program rather than be undertaken concurrently with it. Along with this decision, NASA announced that the huge three-stage *Saturn 5* would substitute for the smaller two-stage *Saturn 1B* as the launch vehicle. This meant the OWS would be sent into orbit "dry" rather than "wet." The S-4B stage, from which the OWS was to be fashioned, would not be employed for propulsion purposes, as was the case when it served as the upper stage of Saturn 1B; rather, *Saturn 5*'s first and second stages would orbit an S-4B already completely outfitted for its orbital lab mission. Since the OWS was to be prepared on the ground rather than in the air, the installation of more elaborate equipment and crew provisions became possible. Also, the greater lifting capability of the *Saturn 5* meant that a larger overall payload could be accommodated. This allowed NASA to include the Apollo telescope mount in this mission.

The name Skylab was made official on February 20, 1970. The first mission included a cluster of four elements: the main OWS, the airlock module (AM), the multiple docking adapter (MDA) and the Apollo telescope mount (ATM).

During this time, Wernher von Braun left his post as director of the Marshall

Center to become deputy associate administrator for planning at NASA headquarters. Taking over von Braun's position was Eberhard Rees, and Leland F. Belew continued as the Marshall Center's Skylab program manager. Mathews became the headquarters' deputy associate administrator for manned space flight, and William Schneider was named the overall director of Skylab. Managing Skylab activities at the Manned Spacecraft Center (later the Johnson Space Center) in Houston was Kenneth Kleinknecht.

The major objectives of the Skylab project were set by NASA. They were: to conduct Earth resources observation; to advance scientific knowledge of the Sun and stars; to study the processing of materials under weight-lessness; and to better understand manned space flight capabilities and basic biomedical processes.

Inside the 118.5-foot long (36.1 meter), 199,750-pound (90,606-kilogram) orbital laboratory NASA intended to conduct 270 scientific and technical investigations. These investigations, which required ninety different pieces of experimental hardware, covered such diverse fields as space physics, stellar and galactic astronomy, solar physics, biological sciences, space medicine, Earth resources and meteorology, materials science and technology, in-space manufacturing and the study of the functioning of man in an orbital laboratory environment.

Nearly 250 participants were involved, including eighty-two foreign

scientists representing forty-five institutions in eighteen countries. In order to comfortably undertake the massive amounts of experiments, the crew was provided with a spacious total living and working area of 12,398 cubic feet (350.86 cubic meters).

With these goals in mind, NASA shot the unmanned Skylab cluster toward its orbit from Kennedy Space Center slightly after noon on May 14, 1973. At first, it was "all systems go" as *Skylab 1* (*SL-1*) streaked through the sky. Everything was working on schedule, and orbital injection was accomplished. The cover that protected the spacecraft from contamination during the launch phase was jettisoned and the ATM was successfully deployed. However, just as *SL-1* turned eastward following the curvature of the Earth and lost contact with ground communications, it was noticed that the deployment signal for the micrometeoroid shield (which deflects small space rocks) had not been received. After fifteen minutes, communications were resumed, but there was still no indication that the shield had locked into position. Even more troublesome was the fact that the "on" light for the OWS's solar array beam deployment remained off. It seemed that the micrometeroid shield, which was also to provide thermal protection, was being inhibited by an airlock. Theoretically, since the solar array beam lay over the top of the undeployed shield, the latter would not be able to lock into position until the beam was extended.

Opposite page: Floating freely is astronaut candidate Anna L. Fisher; in the foreground doing handstands are Dr. Joseph De Gioanni and Dr. Michael A. Berry. This training session prepares those going into space for the effects of weightlessness. *Below:* In the environs of microgravity, Karol J. Bobko and Donald E. Williams juggle various pieces of fruit.

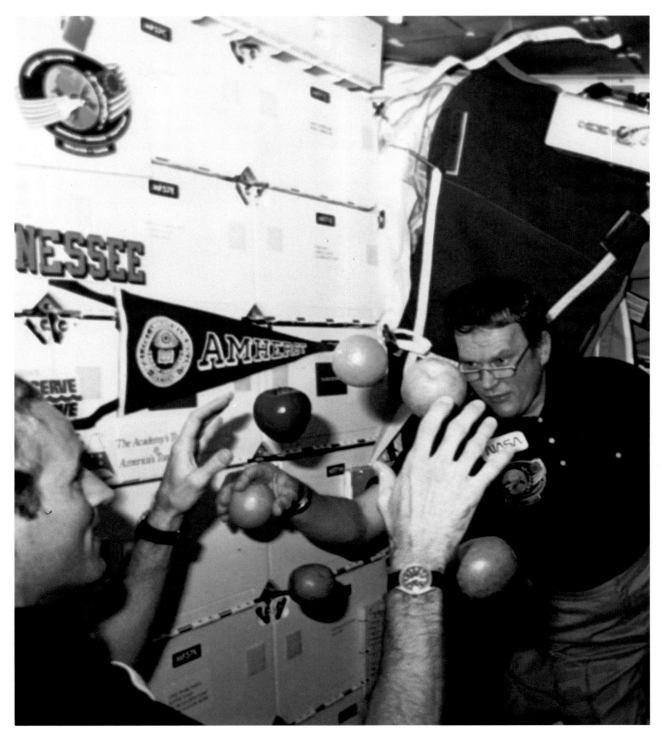

SKYLAB ORBITAL WORKSHOP

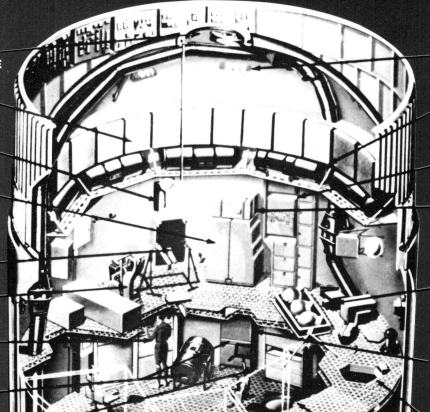

ENTRY HATCH &
AIRLOCK INTERFACE

LOCKER STOWAGE

RADIANT HEATER

FILM VAULTS

ASTRONAUT
MANEUVERING UNITS

M-509 BACK PACK
HAND OPERATED

T-020 SADDLE MODE
FOOT OPERATED

CONTROL
& DISPLAY PANEL

LOWER BODY
NEGATIVE
PRESSURE MO-91

EXP M-171
METABOLIC
ANALYZER

HUMAN VESTIBULAR
FUNCTION M-131

GENERAL
UTILITY LIGHTS

WATER SUPPLY

URINE RETURN
CONTAINERS

ULTRA-VIOLET
AIRGLOW HORIZON
PHOTOGRAPHY EXP

NITROGEN STORAGE
FOR ASTRONAUT
MANEUVERING UNITS

EXPERIMENT SUPPORT
SYSTEM

FORWARD
COMPARTMENT
ACCESS

SHOWER

WASTE

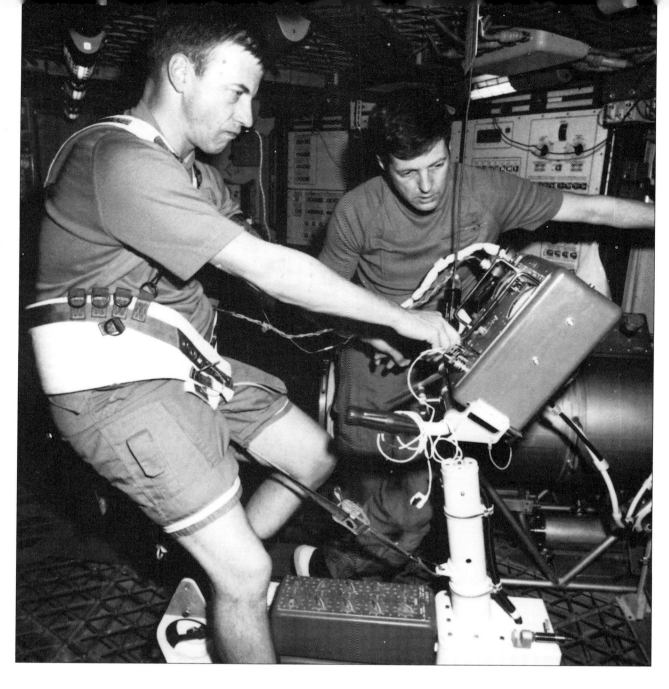

Opposite page: A cut-away of the *Skylab* Orbital Workshop. *Right:* On board the full-sized mock-up of Earth-orbiting *Skylab*, astronaut Paul J. Weitz checks out the bicycle ergometer. The bike was used in an experiment to determine if human capability is diminished or altered by prolonged stays in space. Dr. Kerwin is shown inside the lower level along with astronaut Weitz.

When *Skylab* was in contact with the tracking station near Madrid, Spain, a second solar array beam deployment command was transmitted to the craft. Though the signal was received by the orbital laboratory, nothing happened.

With the failure of both the primary and secondary beam deployments, NASA began to suspect instrument failure and transmitted a new command for shield deployment when *Skylab* came within range of the Hawaiian ground station.

As before, there was no reaction from the troubled spacecraft. Mission control, though baffled, surmised that something had gone wrong with the mechanism for deploying the solar array assembly. During this period, *Saturn 5* launch vehicle personnel reported "strange lateral acceleration" slightly over a minute after liftoff on the S-2 second stage. This occurred when *Skylab* was speeding at just over the speed of sound and seconds before it experienced "Max Q" (the point where the craft undergoes the greatest aerodynamic strain). Soon after, the thermal monitoring crew observed that OWS temperatures were starting to rise out of control. Because the micrometeoroid shield also protected the OWS from the rays of the Sun, engineers concluded that the shield must have been literally ripped off the craft.

This was a problem as far as micrometeor protection (protection from tiny meteors); however, it was not serious. Chances were remote that a particle puncture would occur, causing a loss in OWS pressure during the mission. According to NASA engineers, a tear up to one-quarter inch (one-half centimeter) in diameter could be tolerated and the crew could find it and patch it themselves.

As for the loss of thermal protection, this was another matter entirely; so was the problem of the undeployed OWS solar array panels, which were to furnish at least half of *Skylab*'s electric power. Mission controllers soon realized another flaw in the craft—one of the two solar panels had apparently been torn off and the other panel was blocked in a half-open position.

Outside *Skylab* 3, astronaut Jack R. Lousma participates in the August 6, 1973 EVA, during which he and science pilot Owen K. Garriott deployed the twin pole solar shield to shade the Orbital Workshop. Visible in Lousma's helmet visor is a reflection of the Earth.

UMBRELLA AGAINST THE SUN

The *Skylab* ground crews at Houston, Huntsville and Cape Kennedy now had to ascertain specifically what occurred and what could be done to correct this potentially fatal flaw. The new director of the Marshall Center, Rocco A. Petrone, who replaced the recently retired Rees, stated that "very obviously we had a stranded ship...up there in trouble, [and] we had to make a decision about whether to put men into it and proceed with the mission."

NASA, after much thought, sent up an investigating team whose first task was to reorient the craft in such a way as to diminish the immediate dangers of the Sun's unfiltered heat and thereby lower the interior temperatures to preserve food, film, instruments and other supplies from irretrievable damage.

NASA's solution was to maneuver *Skylab* into a position to stabilize the interior heat without allowing the partially deployed solar panel and the main ATM panels to face away from the Sun, as this would deprive the vehicle of the necessary power. This tedious process took ground engineers two entire days. The method they used was to pitch and roll the vehicle blindly (like parking a boat with remote control outside of the field of vision). This solution, however, was only temporary, because the ground crew could not maneuver the spacecraft for an indefinite period of time. As James Kinsbury, deputy director of Marshall's Astronautics Laboratory, stated, something had to be done to "knock the Sun, the hot Sun, off the vehicle."

Twenty-four-hour work sessions were held at every NASA command center; finally, a long-term plan arose. The first crew would be dispatched toward the craft to insert, through the small scientific airlock in the OWS, a canister that housed a center post and four expandable legs, much like an umbrella. That accomplished, a second crew would extend a couple of poles over the exposed area where the shield had been and then run out a curtain, much like an old-fashioned clothes line. The engineers at NASA were sure this

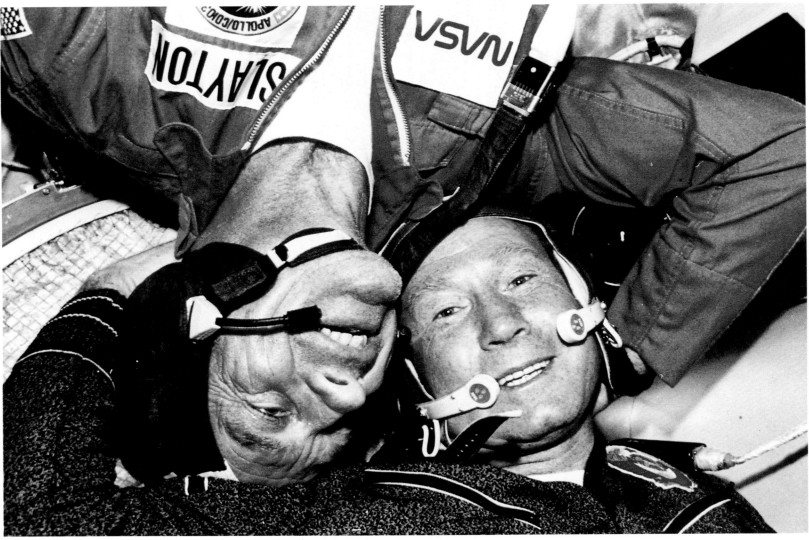

would be possible because they had previously tested it in Marshall's Neutral Buoyancy simulator, a gigantic tank of water in which zero-gravity operations were simulated.

When *Skylab 2* astronauts Captain Charles Conrad Jr., Capital Paul J. Weitz and Doctor Joseph P. Kerwin reached *Skylab* (which was then in a 268.1- by 269.5-mile [432.42- by 434.68-kilometer] orbit, and inclined fifty degrees towards the equator) after their flawless launch on May 25, 1973, they immediately discovered that the micro-meteoroid shield and one solar panel had, indeed, been torn away from the craft. The remaining OWS solar beam was jammed by an aluminum strip that had been attached to the missing shield. After completing a fly-around in their Apollo CSM, the astronauts docked. The following day, they disembarked and climbed into *Skylab*, where their first order of business was to deploy the umbrella shield, blocking out the harmful heat of the Sun. Once this job was finished, Conrad radioed back to Earth, "She's out, and temperatures seem to be coming down."

After the first manned *Skylab* mission, Richard Tousey, a solar astromomer with the Naval Research Lab, said, "I hate to say so but I sometimes think that with these experiments we have recorded more solar information and better solar information than everything we did before, rolled together." Leland Belew said that the experience of *Skylab* "is providing a solid base of factual data and scientific information for use in planning space programs, including the shuttle and advanced space stations . . . Man's performance on *Skylab*, in proving his capability for long duration missions and his ability to save a mission; the successful demonstration of advanced designs in hardware and systems; and the accumula-

Below: On board Spacelab 1, inside the space shuttle Columbia as it orbits the Earth, are its six crew members. Clustered in the aft-end cone are (clockwise): John W. Young, Ulf Marbold, Owen K. Garriott, Brewster H. Shaw Jr., Byron K. Lichtenberg and Robert A. R. Parker. Opposite page: Reviewing their mission plans are the Skylab 4 astronauts (left to right): William R. Pogue, Gerald P. Carr and Dr. Edward G. Gibson.

tion of very high quality data with major scientific significance, all more than fulfill the program expectations." William Schneider, the director of the Skylab missions, added, "We're free [now] to plan on a long-term space station. And eventually this same [experience] will be used to assure ourselves that if we ever want to go out to a distant planet, why that, too, is attainable."

During a three-hour EVA on June 7, Conrad and Kerwin assembled a twenty-five-foot (7.62-meter) aluminum pole on the aft end of the OWS. They attached a cable cutter to one end of the pole and then positioned the pole so that they could clamp cutter jaws onto the jagged strip. They then fashioned the other end of the pole to the ATM truss structure. This "bridge" was used by Conrad to inch his way, hand-over-hand, to the solar panel beam: once there, he held it in place by lying down across it while Kerwin did the physical cutting, closing the jaws of the tool with the assistance of a lanyard (a short cord or rope). Kerwin coolly reported to ground control, "We got the wing out and locked."

A SOLID BASE FOR RESEARCH

Finally, *Skylab* was operational. Weitz, Kerwin and Conrad completed the *Skylab 2* mission and were almost immediately followed, on July 28, by *Skylab 3's* Captain Alan L. Bean, Major Jack R. Lousma and Doctor Owen K. Garriott. The crew of *Skylab 4*—Gerald P. Carr, William R. Pogue and Edward G. Gibson—were launched on November 16 of the same year.

These astronauts, after their respective twenty-eight-, fifty-nine- and eighty-four-day missions, were, once and for all, able to put to rest any lingering misgivings about the benefits of sending men into space. Because of its ability to launch knowledgeable astronauts, NASA was able to preserve a $2.5 billion spacecraft and make key contributions to the world's understanding of the Earth and universe. Once the craft was able to perform, William C. Keathley, Marshall's ATM Experiments Branch chief, exclaimed with a sense of satisfaction that the eight solar instruments on board "worked extremely well, met all of our expectations and ... exceeded those expectations."

Skylab's vast success would definitely have been compromised had the ability of the nine astronauts to perform their duties for lengthy periods of time under zero-gravity conditions been impaired. Fortunately, all the astronauts returned from their missions in space quite healthy. There were, nevertheless, temporary changes in their conditions. There were small downward dips in all of the astronauts' red blood counts and in the calcium content of their bones. Doctors also observed mild irregular heartbeats (known as cardiac arrhythmia) when monitoring the astronauts. However, in every instance, tests illustrated "rapid and stable adaptation to the zero-gravity condition." A few weeks after their return to Earth, all the abnormal readings vanished.

Another observation made by the NASA medical team was that crew members grew from three-quarters of an inch to an inch and three-quarters (one centimeter to four centimeters) taller as the effect of gravity compressing their spinal disks was nullified. Soon after the astronauts returned to Earth, they returned to their smaller stature!

This clear view of the *Skylab* station cluster in Earth orbit was taken from the *Skylab 3* CSM during a "fly around" inspection. Below the craft are clouds and water. Note that the Apollo Telescope Mount has four solar panels deployed.

The *Skylab* astronauts learned that by employing such aids as a stationary bicycle and spring extenders, they could maintain their muscle tone and general sense of physical well-being. This enabled the crew to carry out their work and to keep a positive sense of harmony while aboard the small craft orbiting in the vacuum of space. Pilot Joe Kerwin wrote about his *Skylab 2* mission that "a machine [the astronaut crew] that gets the giggles late at night over the ice cream and strawberries, and occasionally looks out the window when it should be writing the log . . . is replenished by these things, better than a machine is by its voltages and lubricants, and will be ready to go again tomorrow."

What is considered a medical first was conducted during the SL-3 mission where astronauts demonstrated the growth and division of human cells in zero-gravity.

Despite the successful accumulation of scientific data—the discovery that certain industrial processes have their greatest potential in the confines of space, and the observation of such delights as the first recording, from beginning to end, of a solar flare—perhaps the most significant accomplishment of the Skylab missions was the simple knowledge that humans could survive in space for long periods of time without physiological or psychological problems.

"Whenever a distinguished but elderly scientist says that something is possible, he is almost certainly right. Whenever he says that something is impossible, he is almost certainly wrong."

—ARTHUR C. CLARKE

THE SPACE SHUTTLE

With the success of *Skylab*, NASA was looking toward the construction of a permanent space station, which could be a stopping-off point for space traffic to and from Earth. By 1972, spurred on by the success of *Skylab*, NASA began to plan a vastly superior spacecraft. This craft would solve many of the questions that were being posed to the agency. Because of a lack of support from the Nixon Administration, NASA was faced for the first time with the need to create a cost-efficient program, as well as one that could utilize reusable parts. To answer these and countless other questions, NASA introduced the Space Transportation System (STS), or the Space Shuttle as it is commonly known, to replace all of the then current and expendable launch vehicles.

LOW-COST ACCESS TO SPACE

After gaining approval from President Nixon and Congress, NASA made the STS its major priority and began to alter its traditional view—one that saw space activities as idiosyncratic activities of and for a limited few—to the more forward-thinking perspective that saw space activities as an integral part of human activity and a tremendous benefit to all of society. NASA believed that the more space became an everyday part of life, the more beneficial the Space Transportation System would be. The STS was designed to provide low-cost, reliable transportation from the Earth's surface to low Earth orbit and return.

The revised program had four components:

• Development of space transportation

capabilities—acquisition, testing, production and continued improvement of space vehicles and the services they provide.

- Space flight operations—prelaunch, launch, flight, landings, post-landing activities and the related customer services.

- Active support of the commercial use of space and of privatizing space transportation systems such as expendable launch vehicles and upper stages for the Shuttle.

- Advanced programs—planning and evolutionary development of follow-on programs to exploit the STS, define a second-generation system, provide an infrastructure for permanent presence of humans in space and increase space flight capabilities through development of advanced transportation, satellite services, advanced crew and life support and tethered (tied) systems.

- The Shuttle's major advantage is its ability to service, maintain, repair, retrieve and reuse payloads. Its most important feature is its versatility, which will make utilizing all of its possibilities a constant challenge.

Opposite page, above: At Kennedy Space Center, a worker examines specially coated black tiles, called High Temperature Reusable Surface Insulation, as the orbiter Columbia is inspected. Close to 75 percent of the Shuttle is shielded from heat by the approximately 31,000 individual tiles. Opposite page, below: At the Wabash, Utah, facility of the Morton Thiokol Corporation, a motor is static tested. Morton Thiokol is NASA's prime contractor for STS solid rocket motors. Left: Enroute from the Vehicle Assembly Building to the orbiter Processing Facility is the space shuttle Atlantis. At the OPF the craft will begin extensive preparations for its next mission.

INSIDE THE SHUTTLE

Empty, the space shuttle orbiter has a weight of about seventy tons (seventy-seven metric tons). When the maximum payload weight of about 60,000 pounds (27,216 kilograms) is included, the total weight placed in orbit on each flight is 100 tons (110.2 metric tons). If the external tank (ET) were carried along, the total weight in orbit would actually be increased by close to two tons. The size of the overall booster is set by the ability to orbit 140 tons (154 metric tons) at a time, more than equal to the payload capacity of the old *Saturn 5.* The Shuttle can leave only about twenty percent of this weight in space; eighty percent of the mass and cost of each launch is used up launching the orbiter and the ET.

The Shuttle itself, a winged, maneuverable, aerodynamically sleek spacecraft, is powered by a set of three hydrogen-oxygen engines that represent the pinnacle of space design. They are shielded during reentry by thousands of high-performance heat shield tiles. The Shuttle crafts are launched from the pad by two huge recoverable, solid-propellant strap-on rocket boosters (SRB's), and draw their supply of liquid hydrogen and oxygen from an even more immense external tank that rides along most of the way into orbit. The shuttle performs an additional maneuver to jettison the ET into the ocean rather than carry it into orbit. Once it is detached, the ET is retrieved and reused.

Of the shuttle's three main elements—a space plane (the orbiter), two rocket boosters, an external tank—only the tank needs to be replaced for each mission. A popular way of perceiving this particular craft is as a space "truck" capable of delivering cargo, such as a satellite or Spacelab, into orbit and then returning to Earth for another mission.

Inside the orbiter is the forward fuselage, which has two levels. The upper level serves as the flight deck, where a crew of seven are stationed at control panels. In the forefront of the flight deck is a cockpit, where the flight pilot and the mission commander sit. The lower level, or middeck, serves as the living area for the astronauts. There the men and women can wash, eat, relax or

Far left: A redesigned engineering test motor is successfully static tested at the Wabash, Utah facility of Morton Thiokol. This new motor no longer has the third "O" ring—a change brought about by the explosion of the *Challenger. Left, top:* Carefully, the orbiter *Atlantis* is hoisted from the shuttle stack to the transfer aisle during de-stack operations. After this, the craft will be transferred to the Orbiter Processing Facility. *Left, middle:* As preparations for de-stack begin, the shuttle *Atlantis* is transferred from the Vehicle Assembly Building high bay 1 to high bay 3. *Left, bottom:* In the Vehicle Assembly Building, preparations commence for de-stacking of the *Atlantis* from the external tank and the solid rocket boosters.

sleep. Located behind the middeck is the cargo bay—a huge compartment measuring sixty feet (19.29 meters) long and fifteen feet (4.57 meters) wide. Astronauts must wear pressurized suits when entering and working in this area because, unlike the front fuselage, it is not pressurized.

Underneath the cargo bay are sev-eral liquid hydrogen and oxygen tanks that supply the fuel for the on-board electrical power system. In the tail end of the orbiter the three main engines, which also burn liquid hydrogen and oxygen, are clustered. Also located there are the twin engines of the Orbital Maneuvering System (OMS) that pro-pel the orbiter into Earth orbit and, later,

At Kennedy Space Center, the space shuttle *Discovery* begins its move toward launch pad A, from where it made its maiden flight.

help to slow down the craft as it reenters the Earth's atmosphere.

The orbiter itself is mounted on the side of the huge external tank, to which are strapped the two Solid Rocket Boosters. The boosters, 149.16 feet (45.46 meters) high and 1,214 feet (3.7 meters) in diameter, and the external tank, 154.2 feet (47 meters) high and 27.56 feet (8.4 meters) in diameter, tower over the orbiter.

Typically, the Shuttle blasts off from the launch pad at the Kennedy Space Center. There, it uses the same pads that launched the Apollo missions. The brightly colored exhaust seen in photographs and videotapes comes from the two Solid Rocket Boosters (SRB's)

Far left: The space shuttle *Atlantis* soars through the blue Florida sky on its maiden flight. *Atlantis* was the fourth shuttle to be launched by NASA. It carried a five-man crew and a classified Defense Department payload. *Left, top:* At Kennedy Space Center, the shuttle *Challenger* sits atop the Crawler Transporter on its way to the launch pad. *Left, middle:* As the shuttle *Discovery* is fired up, launch personnel monitor the status of the orbiter. *Left, bottom:* This is the second roll-out for the *Columbia*, as it arrives at the launch pad in the hazy light of dawn. A nozzle on the solid rocket booster had to be replaced before NASA officials would permit the launching of *Spacelab 1*, which carried the first European to fly in a United States spacecraft.

A beautiful view of the Finger Lakes area of New York, including Lake Ontario and a tiny section of Ontario, Canada. Along with the view of remnants of the Glacial Bar, the cities of Rochester, Syracuse, and Utica, New York can be seen.

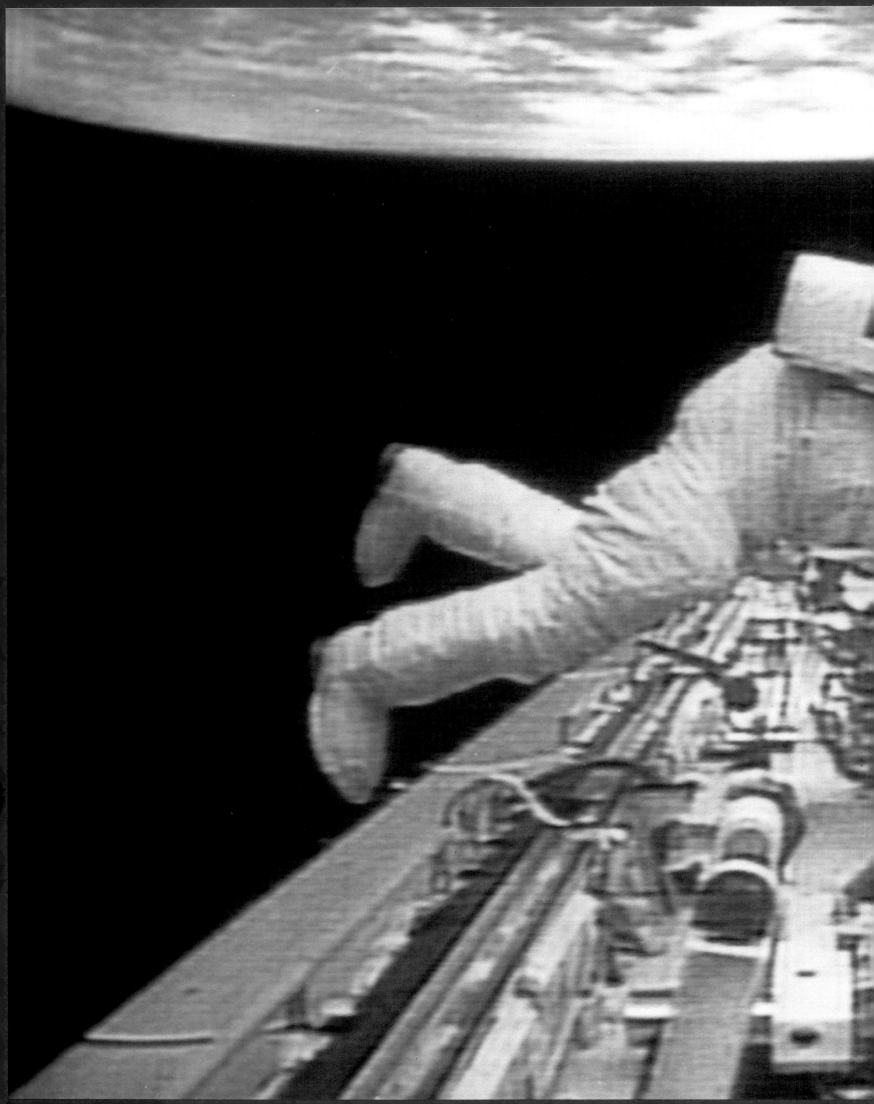

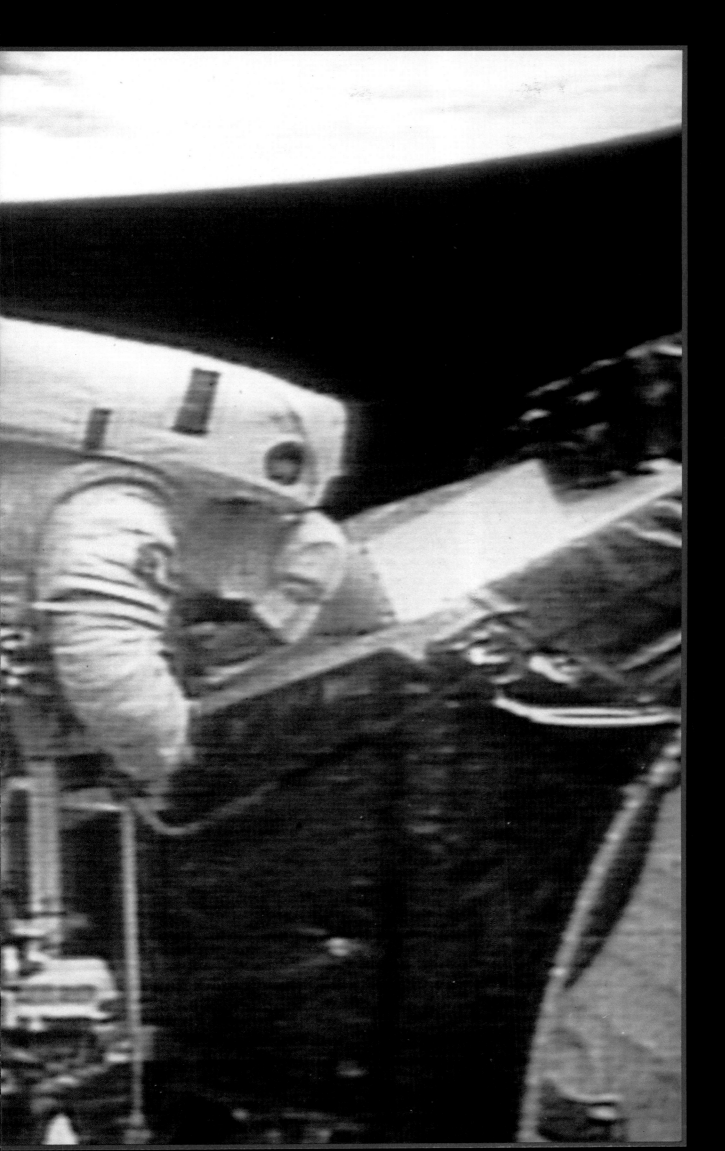

The Earth's horizon can be seen at the top of this frame, recorded from a live TV transmission on April 7, 1983. Astronaut Donald P. Peterson, the mission specialist, busies himself near the airborne support equipment during EVA on day four of the *Challenger's* inaugural flight in space.

attached to the gigantic External Tank (ET). The SRB's burn for about two minutes and are then dropped and parachuted into the Atlantic Ocean to be recovered for later use. The three main Shuttle engines on the tail of the orbiter burn the hydrogen and oxygen from the ET tanks—the flame then is virtually invisible, due to the fact that it contains no carbon compounds or metals that produce such luminous particles. The main engines burn from liftoff to the point of jettisoning the ET, just before orbital injection. Maneuvering in space by the orbiter is accomplished with two much smaller Orbital Maneuvering System (OMS) engines. These engines are then fed by the tanks of stored liquid propellant carried in the orbiter's tail. At this point, the craft's speed is approximately 17,500 miles (28,000 kilometers) per hour. The shuttle reaches an altitude of 150 to 200 miles (240 to 320 kilometers) and begins its continuous orbit

around the Earth. A standard orbit takes only one and a half hours.

To reenter the Earth's atmosphere, the orbiter must brake to below orbital velocity so that the Earth's gravity can capture it and begin to slow it down. The orbiter turns around so that its tail faces forward, and the OMS rockets are then fired against the flight direction. The orbiter then continues its descent as a glider, entering the atmosphere at twenty-five times the speed of sound. Slowly it loses speed and again rotates. At 1,700 feet (518.16 meters) the orbiter's nose tilts upward and begins to descend at an angle seven times sharper than that of a jetliner. At 900 feet (274.32 meters), the landing gear extends, and in less than half a minute it lands horizontally on its three-mile (4.8 kilometer) runway at Kennedy Space Center.

The Shuttle's ablative tiles habitually chip or fall off; the strong, large Shuttle

A Solar Maximum Mission Satellite awaits its return to service in Earth orbit in this wide-angle scene of the *Challenger's* cargo bay. The remote manipulator system (RMS) seen at the right edge was used to capture the previously dormant satellite and later, to return it to space. *Challenger* was flying at 285 nautical miles above the Earth when astronaut Dick Scobee took this picture.

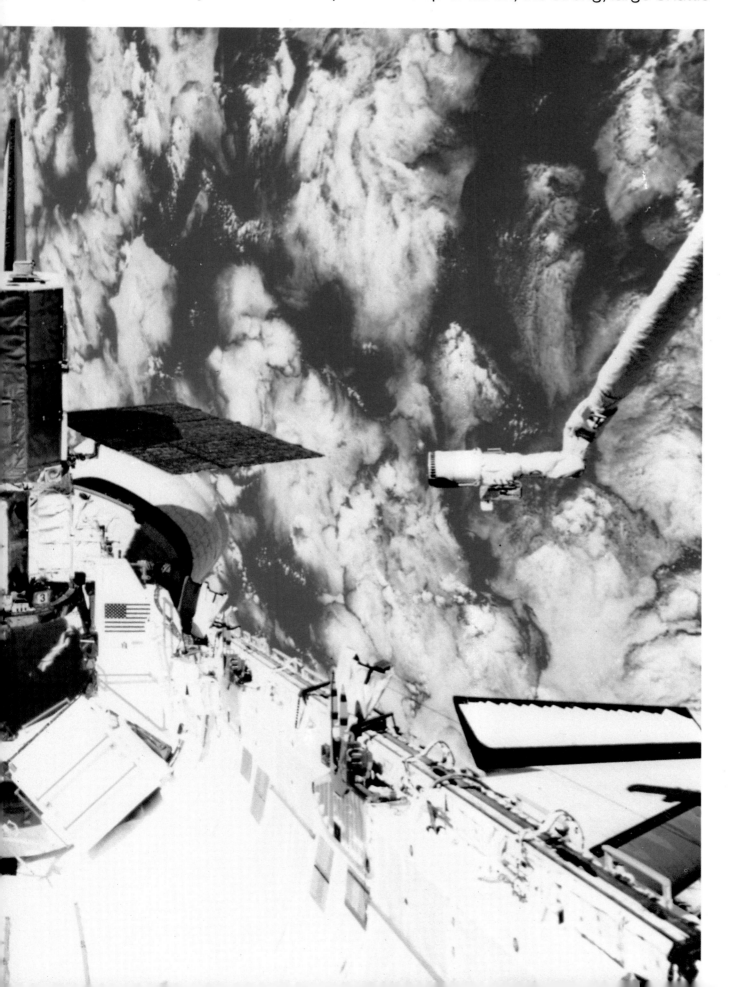

Right: Astronaut Bryan O'Connor photographed this scene of Jerry Ross's EVA egress with a 35mm camera and a 16mm lens. In the foreground are part of the airlock and its hatch. *Opposite page, top to bottom:* An interior view of the shuttle mission simulator (SMS) in the mission simulation and training facility at the Johnson Space Center. A variety of displays and controls are seen in the photograph, including the cathode ray tube, computer displays, and the commander and pilot computer keyboards. On board the shuttle *Discovery,* astronaut Judith Resnick, mission specialist, anchors herself on the flight deck. Prior to their EVA designed to deploy snagging devices, Karol Bibko, mission commander, assists Jeffrey A. Hoffman. Astronaut Joe H. Engle, a Kansas native, holds a sigh with a memorable line from the "Wizard of Oz." As part of the extensive medical tests conducted on board, United States Senator E. J. (Jake) Garn conducts an experiment on himself.

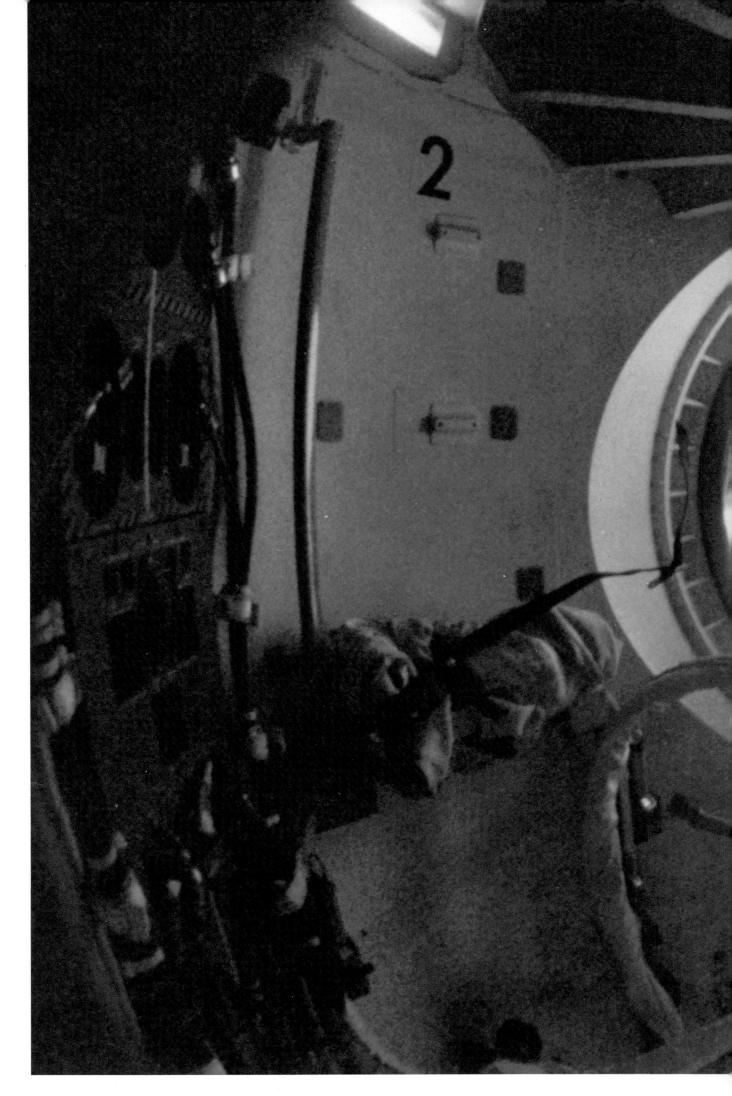

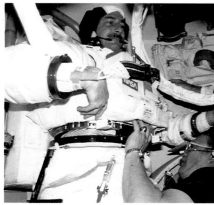

TOTO...
I DON'T THINK WE'RE
IN KANSAS ANYMORE

Right: This photograph taken on board the Shuttle Pallet Satellite (built by a West German firm) during the *Challenger* flight on June 22, 1983, in Earth orbit, shows the Canadian-built remote manipulator system (RMS), the experiment pallet for NASA's Office of Space Terrestrial Applications, the closed protective "cradle" device for the now vacated Teles-tat Anik C2 satellite and several getaway special canisters (GAS Cans). *Below:* Astronauts George Nelson (right) and James van Hoften use the mobile foot restraint and the RMS as a "cherry picker" device for moving the "captured" Solar Maximum Mission satellite (SMMS) in the aft end of the Challenger's cargo bay. Later, the RMS again lifted the SMMS into space. *Opposite page, above:* On board the *Challenger,* mission specialist astronaut James van Hoften finishes a busy day with his first "field" application of the Manned Maneuvering Unit (MMU), a nitrogen-propelled, hand-controlled backpack device. At the photograph's edge is the "forearm, wrist and end effector" of the RMS. *Opposite page, below:* Astronaut George Nelson makes the first excursion to the damaged Solar Maximum Mission Satellite (SMMS) on April 8, 1984.

wings must be shielded with thermal protection tiles to safeguard them from burning off; and the retrievable SRB's crash into the ocean so powerfully that they are often damaged beyond expectation. But the fact remains that these are not permanent obstacles to using the Shuttle system; rather, they are the problems of adjusting to a more sophisticated program.

The Shuttle program is tremendously expensive, costing billions of dollars more than anyone anticipated. Why? A major reason for the expense is that, unlike a jet plane, the Shuttle's launch preparations are extensive. A Shuttle launch requires the manpower of a massive crew to mount it vertically on the mobile launch platform, located on the back of an oversized crawler transporter that positions the whole unit over a blast pit. The solid rocket

boosters are the reassembled pieces previously retrieved from the ocean. This loading, launching, monitoring and retrieving is an extremely complicated, time-consuming and expensive process. It is important to remember that although the Shuttle has the appearance of an airplane, it is actually the upper stage of a modern rocket.

The Soviet Union has a shuttle program of its own, although little is known about it outside of Russia. It is very similar to the NASA shuttle in size and design—from its swept-back wings to its 38,000 heat-resistant tiles—although it does differ in many important details. The main engines are not reusable and are located on the external fuel tank. It is unclear how the Soviets plan to use their shuttle system, although it is assumed that it will carry supplies to and from the space station.

Far left: Astronauts Dale Gardner (left) and Joseph Allen work together with Anna Fisher (not in picture, but controlling the RMS arm from *Discovery*'s cabin) to bring Westar VI into the Shuttle's cargo bay for repairs. Allen is on the mobile foot restraint, which is attached to the RMS end effector, while Gardner removes a stinger device from the now stabilized satellite. *Left:* Near the end of a long-duration task during the retrieval EVA, astronaut Joseph Allen IV appears to be hoisting huge weights while astronaut Dale Gardner holds on. Dr. Allen is actually the only anchor for most of the "captured" Palapa B-2 communications satellite. Gardner used a torque wrench to secure the satellite to its "parking place" in the shuttle's cargo bay. *Below:* Approaching a hard dock with the spinning Westar VI satellite over Bahama banks, is astronaut Dale Gardner, wearing the MMU. The RMS arm, controlled by Dr. Anna Fisher inside *Discovery*'s cabin, awaits its turn.

Far left: Astronaut McCandless uses the combination RMS arm and mobile foot restraint to experiment with the "cherry picker" idea. *Left:* During *Challenger*'s fourth flight into space, astronaut Robert Stewart glides several meters above the cargo bay while conducting this flight's second EVA. *Below:* Astronauts Ross (left) and Spring (bottom) at work during their December 1, 1985 EVA.

Astronaut Jerry Ross, anchored to the foot restraint of the RMS, does work on ACCESS. This photograph was taken by astronaut Sherwood Spring during the week-long 1985 mission.

Below, top: Guion S. Bluford, NASA's first black astronaut. *Below, bottom:* Dr. Sally K. Ride, NASA's first woman in space. *Right:* The space shuttle *Atlantis* about to land on the dry lake bed at Edwards Air Force Base, California, on October 7, 1985. This mission carried commander Karol Bobko, pilot Ronald Grabe, mission specialist David Hilmers, Robert Steward and Air Force Major William A. Pai.

SHUTTLE BENEFITS

The first orbiter, the *Enterprise* (rumored to be named as a salute to the television series *Star Trek*), made its initial test voyage on August 12, 1977, with the launch assistance of a Boeing 747. With astronauts Fred Haise Jr. and Lieutenant Colonel Charles Fullerton aboard the *Enterprise*, the 747 took the orbiter up on its back and then released it at an elevation of 22,800 feet (6,949 meters). About one month later, on September 13, the *Enterprise* was again launched, piloted by Colonel Joe Engle and Commander Richard Truly. After its fifth and final test flight using a Boeing 747, NASA delayed its scheduled launches due to the loss of heat shield tiles.

In May 1979, NASA resumed testing and the *Enterprise* was once again wheeled out of the Vechicle Assembly Building. On April 12, 1981, NASA launched the first orbital test flight of the Space Shuttle *Columbia*, with John W. Young serving as commander and Captain Robert L. Crippen acting as pilot.

There were four major orbiters: *Columbia, Discovery, Atlantis* and the ill-fated *Challenger*. Throughout the ensuing years each of these crafts performed many tasks and gave the United States its first woman in space when, in 1983, Doctor Sally K. Ride orbited Earth aboard the *Challenger* from June 18-24. The nation's first black astronaut, Lieutenant Colonel Guion Bluford, orbited the Earth from August 30 through September 6, 1983. In November of the same year, the initial flight of the Spacelab orbital laboratory was launched aboard the shuttle *Columbia*.

Mission Specialist Bruce McCandless II, aboard the *Challenger*, made a maiden flight into orbit. On August 27, President Reagan announced that the first layperson in space would be an educator, and NASA began to accept thousands of applications. In July 1985, two space "firsts" were announced: The first made-in-space product, polystyrene beads, was made available on the commercial market, and Christa McAuliffe was chosen by NASA from thousands of applicants to be the first "average citizen" to enter space.

This photograph shows the space shuttle *Challenger* returning to Kennedy Space Center atop the 747 Shuttle carrier aircraft. The mission ended at Edwards Air Force Base in California and the duo returned to Florida one day later.

On January 28, 1986, the Space Shuttle *Challenger* exploded just seventy-two seconds after liftoff, killing all the crew members: Michael J. Smith, Francis R. Scobee, Ronald E. McNair, Ellison S. Onizuka, Gregory Jarvis, Judith Resnick and America's first teacher in space, Christa McAuliffe. This accident marked the first human fatalities during a NASA mission. The fatal flaw was soon traced to faulty "O" rings on the solid rocket boosters. These rings separate the fuel sections.

Despite the fact that the *Challenger* accident temporarily closed the gate on low-cost access to space, the Space Shuttle did provide many opportunities for space access, and a wide range of people have shown continued interest in this access.

This crew portrait was taken at Johnson Space Center, Houston, Texas, prior to the ill-fated January 28, 1986 *Challenger* mission.

Right: Technicians prepare *Spacelab 1* for being hoisted out of its test stand and into the payload transport canister. The module was built by the European Space Agency. *Opposite page, left, top:* This frame contains only some of the 3,000 bees which were carried on board. The colony remained in space for the seven day mission. The idea was devised by then-high school senior, Dan Poskevich of Waverly, Texas. *Opposite page, left bottom:* Post-flight photographs of pine seedlings flown on *Spacelab 2.* Plant growth in zero-gravity interests NASA scientists because it may be more cost-efficient for space colonies to grow food in space than to transport it from Earth. *Opposite page, right top:* This *Spacelab 3* mission begins with the lift-off on board the shuttle *Challenger. Opposite page, right, below:* Judith Resnick, mission specialist, controls the unfolding of the solar Array Experiment (SAE) developed for this *Discovery* flight.

GAS CANS IN SPACE

One of the most popular opportunities has been the Get Away Special, known as a GAS can. It gets its name partially from the acronym and partially due to the shape of the payload. A GAS can is a cylindrical can about forty inches (102.5 centimeters) in diameter and either fourteen or twenty-eight inches (35.9 or 71.8 centimeters) high, used for ferrying experiments into space. GAS cans are stored in the side of the Shuttle's cargo bay. If you utilize a GAS can, the astronauts turn on your instrument and then operate it by flicking two toggle switches during flight. You are requested to deliver your instrument to NASA two to three months before the scheduled launch, and your package is returned two weeks after the shuttle lands. The cost and the paperwork are minimal; the prices quoted are from $3,000 to $10,000. The package must pass a NASA safety inspection.

Customers of the GAS can program

are enormously varied. GTE has flown several experiments on light bulbs in GAS cans. High school and college students have designed experiments that have flown as GAS cans. A popular seed company flew twenty-five pounds of herb, vegetable and flower seeds in a GAS can. Astronomers studying the ultraviolet background of the Sun loaded two adjacent GAS cans with telescopes and then placed the controlling avionics in an adjacent third canister. The cluster of three cans flew successfully in an early flight in 1986.

Before the *Challenger* accident, close to fifty GAS can experiments were planned per year. These experiments take up very little room and their weight is considered minimal.

It is essential that scientists, especially space scientists in training, have the opportunity for such low-cost access to space. NASA has, over the past two decades, run a rocket program that permits graduate students a chance to fly an experiment at a low cost. Well-established scientists have also utilized the rocket program to test out an idea relatively quickly, without expensive risk.

With the Shuttle program, NASA developed a sophisticated method of hauling these experiments into space. The Spartan Project, for example, is a box that is deployed as a sub-satellite of the Space Shuttle and is then retrieved after a mission of several days. The last *Challenger* flight was carrying a Spartan project.

Another, even more sophisticated use of the Shuttle is the Spacelab program. This program enables independent contractors to send science payloads into space without building and flying an independent satellite. In a Spacelab mission, the Shuttle's cargo bay is filled with a combination of pallets, which are exposed to the environment of space, as well as pressurized cars in which the astronauts can move about. The pallets are used by astronomers and space scientists for experiments. Materials-processing and Earth-watching experiments are typically stored in the pressurized modules. Before 1986, four Spacelab flights were flown. Three more are scheduled to be launched before 1990.

As the Space Shuttle system slowly moves back into use, NASA scientists and private industry will look toward the Shuttle to continue this type of opportunity in even greater numbers. With the construction of the Space Station imminent, the Space Transportation System will again assist in launching payloads into orbit and to carry parts of the Station into orbit, to be assembled later by astronauts aboard another shuttle craft.

In the years to come, those pioneers who continue with the program through the clouded period of doubt will reap the benefits of this unique access to space.

GOOD MORNING, DISCOVERY

On September 30, 1988 the Space Shuttle *Discovery* was launched from Kennedy Space Center in Cape Canaveral, Florida, signaling the United States' return to the exploration of the stars. The five man crew consisted of: flight commander Captain Frederick H. Hauck who had flown on two previous shuttle flights, the *Challenger* mission on June 18, 1983 and the *Discovery* mission on November 8, 1984; Colonel Richard O. Covey who had flown on the *Discovery* on August 27, 1985; mission specialist Michael (Mike) Lounge who had previously flown on the August 27, 1985 *Discovery* mission; mission specialist Lieutenant Colonel David C. Hilmers, who had flown on the *Atlantis* flight in February, 1985; and mission specialist George D. (Pinky) Nelson who had flown twice before—on the April 6, 1984 *Challenger* mission and the January 12, 1986 *Columbia* mission. While in space, the crew deployed a Tracking and Data Relay Satellite System (TDRS-2) and a 5,000 pound (2,272 kilogram) communications satellite almost identical to the one destroyed in the *Challenger* disaster. The TDRS-2 is an important step towards a system which will eventually replace most of NASA's worldwide network of old and costly ground stations. The TDRS-2 will handle multiple frequencies and will receive and transmit in seconds tremendous amounts of information to and from the shuttle and other spacecraft (such as the soon-to-be-launched Hubble Space Telescope and the United States Space Station to be built in the next decade), as well as dump, or funnel, messages to the NASA ground station in White Sands, New Mexico.

The TDRS-2, with its five-story solar arrays, is the world's largest and most complex communications satellite. It will stay in geosynchronous (stationary) orbit, 22,000 miles (35,500 kilometers) above Earth.

After the TDRS-2 was orbited, the crew conducted a thorough check of the 200-plus changes made to the craft since 1986.

Four days later, on October 3, 1988, *Discovery* successfully landed at Edwards Air Force Base in California.

The space shuttle *Discovery* is rolled out to its launch pad as preparations for its mission continue at Kennedy Space Center.

Opposite page: Newly designed after the *Challenger* disaster, the space shuttle *Discovery* begins its historic lift off from Cape Canaveral, Florida on September 30, 1988. *Right:* The astronaut crew insignia for this shuttle mission include the names of the crew, whose portraits are below. They are, *clockwise from top:* flight commander Captain Fredrick Hauck, mission specialist Mike Lounge, mission specialist Lieutenant Colonel David Hilmers, mission specialist George Nelson, and Colonel Dick Covey.

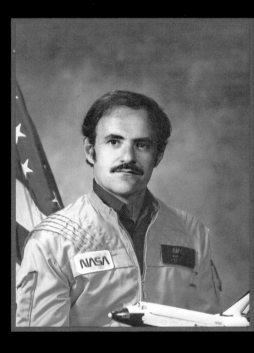

Astronaut Bruce McCandless II nears the maximum distance he can travel from the Earth-orbiting shuttle *Challenger.* McCandless is in the midst of the first "field" tryout of the nitrogen-propelled MMU.

CHAPTER 6

"We have taken to the Moon the wealth of this nation . . . We have brought back rocks, and I think it's a fair trade. For just as the Rosetta Stone revealed the language of ancient Egypt, so may these rocks unlock the mystery of the origin of the Moon, and indeed even of our Earth and solar system."

—ASTRONAUT MICHAEL COLLINS
IN A SPEECH TO A JOINT SESSION
OF CONGRESS, SEPTEMBER 16, 1969

THE FUTURE IN SPACE

Exploration of the solar system has spanned more than two decades and has produced a virtual avalanche of discoveries and a storehouse of data. In the past twenty years more than twenty-four unmanned spacecrafts have transformed our view of the planets from distant telescopic images to clear, concise global perspectives.

During the twenty years from the first *Mariner* fly-by of Venus to the second *Voyager* encounter with Saturn, robot fly-by crafts have visited every planet known by the original astronomers. Most of those crafts, with names like *Ranger, Surveyor, Pioneer, Mariner, Voyager* and *Viking*, belonged to the

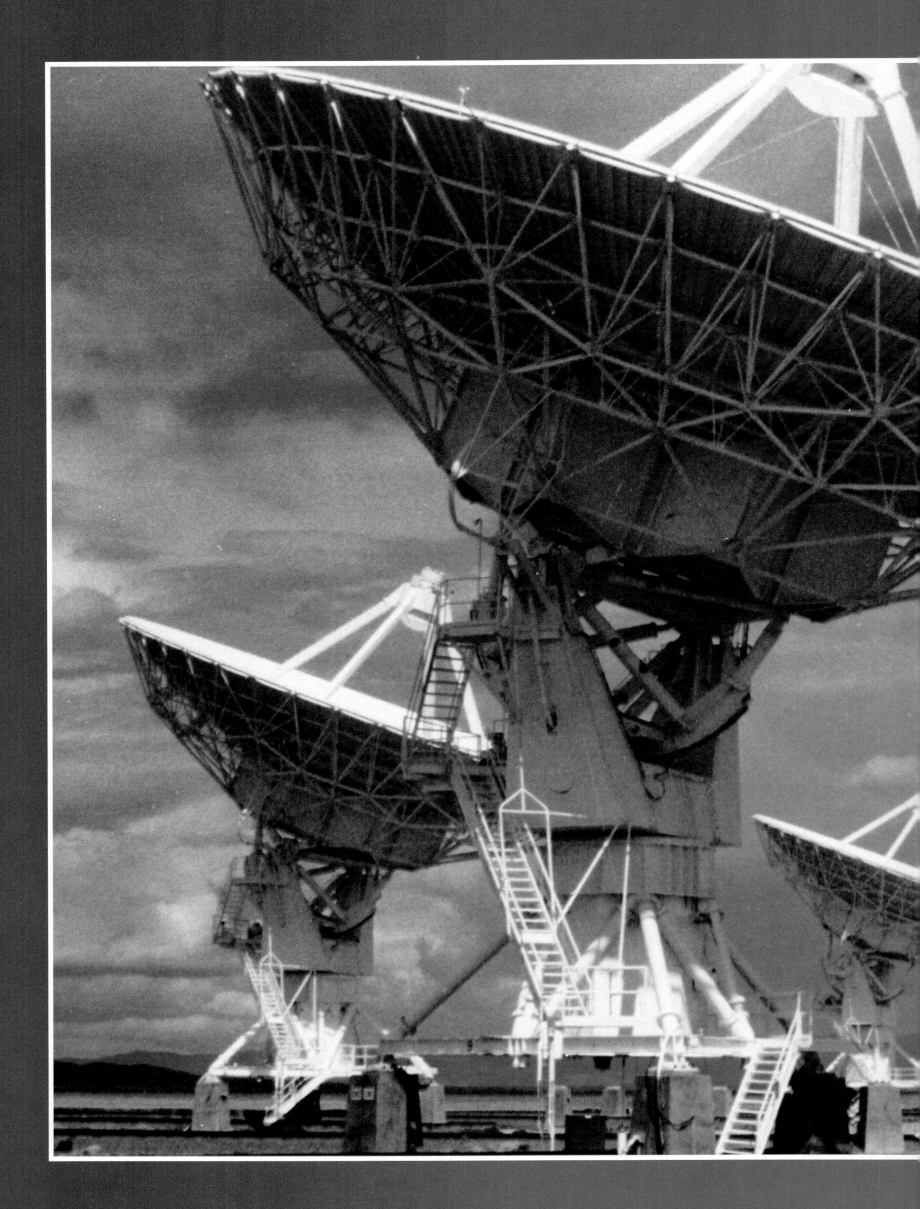

Left: One of twenty-seven, eighty-two-foot (twenty-five-meter) diameter radio telescope antenna dishes at the Very Large Array near Socorro, New Mexico. The VLA is the largest effective radio telescope in the world. *Inset, top to bottom:* In the 1970s, NASA's Planetary Programs spacecraft included: The Mariner/Mars Orbiter, which orbited Mars in 1971. The *Viking,* which orbited and landed on Mars in 1976. The *Pioneer,* a Jupiter flyby probe. *Helios,* an interplanetary probe, which was operational in 1974 and 1975 The Venus/Mercury Flyby.

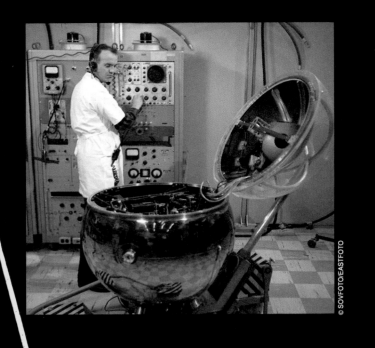

Clockwise from right: Technicians prepare the *Venera 15* for mechanical testing.

GOES-C (Geostationary Operational Environmental Satellite) is being enclosed inside its payload aboard a Delta rocket.

A test model of NASA's *Galileo* spacecraft at the Jet Propulsion Lab. Scheduled for launch in 1990, the craft, twenty-five feet (seven meters) tall, will orbit Jupiter for about twenty months and transmit an instrumented probe into the giant planet's atmosphere.

On board a *Delta* rocket from the Western Test Range in Lompoc, California, *Explorer* is readied for launched into Polar orbit. It studied important energy transfer—atomic and molecular processes that occur in the Earth's upper atmosphere and are critical to maintaining the heat balance of our atmosphere.

A 400-pound (182-kilogram) steel satellite undergoes a final check before being mated to its *Delta* launch vehicle. It will later be placed 155 miles (248 kilometers) up into orbit to measure the structure of Earth's upper atmosphere.

Ignition stages 2-4

Burnout stage one

Stage 3

Stage 2

Stage 1

70 miles

Tracker AMR

Cape Canaveral

Puerto Rico

1500 miles

Earth's rotation 15 deg/hr

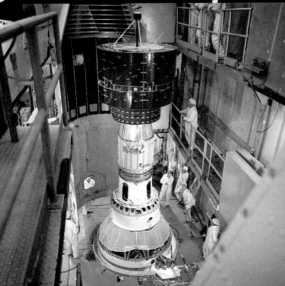

To earth's

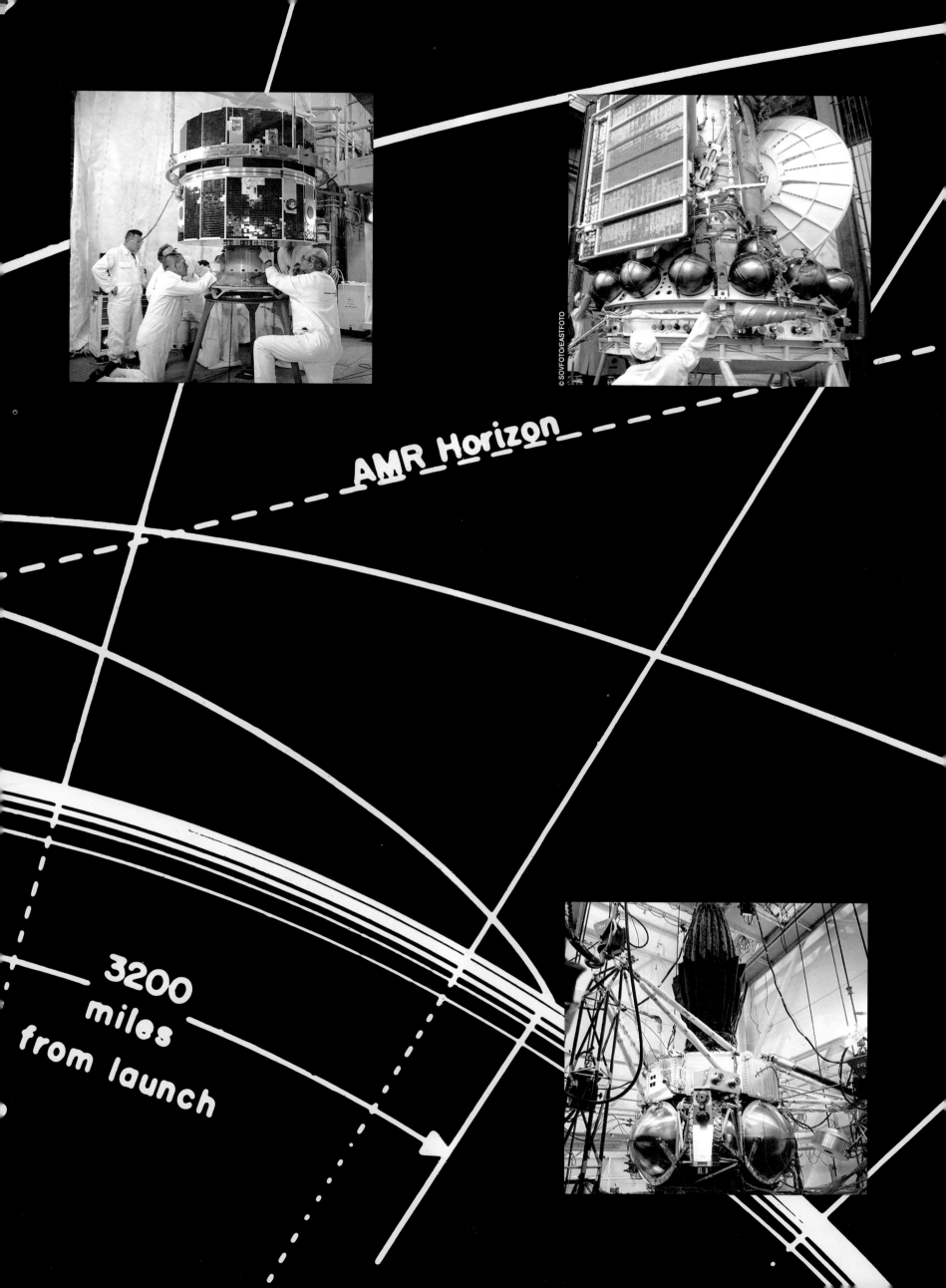

AMR Horizon

© SOVFOTO/EASTFOTO

3200
miles
from launch

United States. The Soviet Union, concentrating on the Moon, Mars, and Venus, also greatly contributed to this era of discovery. Because of this exploration, more than forty planets and satellites of two ring systems have been closely surveyed within the span of a single generation.

Our exploration of the planets represents a resounding triumph of imagination and will for the entire human race. Generations to come may, perhaps, better appreciate the vast scientific significance of these voyages, which we have now come to regard as almost commonplace occurrences.

The approach taken to the exploration of space has been the opposite of how Earth-bound explorations were conducted. From space, we begin with

a global perspective and later shift to a more detailed series of observations and measurements of a specific region. This approach, coupled with what scientists have gleaned about Earth, has become an awesome tool and accounts for the rapid progress being made in the space sciences; however, experience has also shown that vague theories can become overly optimistic without *in situ* (on-site) evaluation. It has become apparent that data based on reconnaissance missions, such as the lunar expeditions, requires a continuing analysis. Therefore, we know that any future visits to planets already explored will encompass, inevitably, many more surprises. This type of exploration will without a doubt, continue into the next century.

Right: Hawaiian Island Wakes. From the Space Shuttle, a simultaneous photograph paired with another of the island chain records only the non-polarized light reflected from the ocean and the earth. The view above clearly shows subtle changes in the water caused by light reflected from beneath the surface by suspended sediments and biological activity. Such a pair of polarization images allows Earth scientists to see both surface and subsurface oceanographic features. Further studies will be done by digitizing each photograph and using computer enhancement techniques to extract other details.

Far left, above: A photograph of a 3987 pound (176 kilogram) paddle wheel spacecraft in a simulated space atmosphere. This is

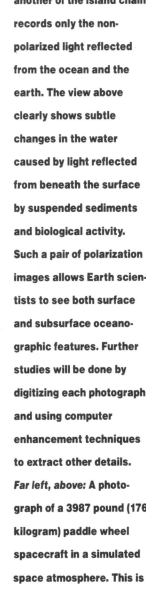

one of several tests for controlling and maneuvering spacecrafts from Earth. *Far left, below:* Final checkout and inspections are made by technicians of the *Mariner 9* spacecraft prior to its being encased and mated with its launch vehicle, the *Atlas-Centaur.* The *Mariner 9* was successfully launched on May 30, 1971. *Right, above:* A photograph showing the mating of spacecraft for ITOS-E, a United States weather satellite which was launched on July 16, 1973 from NASA's western test range. *Right, below:* A development model of a *Voyager* craft is shown at Kennedy Space Center during transport and rocket shroud encapsulation tests. *Voyager* was developed for NASA by the Jet Propulsion Lab in Pasadena, California.

THE SOLAR SYSTEM EXPLORATION COMMITTEE

In order to realistically fashion a comprehensive and flexible space exploration program, the NASA advisory panel established the Solar System Exploratory Committee (SSEC) in 1980. The SSEC received a mandate to develop a mission strategy for solar system exploration through the end of this century.

NASA initiated the SSEC to pose important basic questions about the origin of our solar system that could, in turn, be answered through exploration. The SSEC reached the following conclusions with respect to the planetary exploration program:

"In order to maintain U.S. leadership in solar system exploration and to realize any reasonable progress toward the scientific goals recommended by the Space Science Board, NASA should immediately initiate the Core planetary program outlined below. The Committee also urges that this Core program be augmented at the earliest opportunity with additional, more technologically challenging missions of high scientific priority and exploration content."

The Core program will have two facets: ongoing basic activities, including basic research, missions operations, technology development and advanced planning; and the Core "planetary missions program." If the total budget level of $300 million per fiscal year is to be maintained for this program, it is inherent that an innovative approach to spacecraft mission design is developed. The initial Core missions recommended by the SSEC are the Venus Radar Mapper, the Mars Geoscience/Climatology Orbiter, the Comet Rendezvous/Asteroid Flyby and the Titan Probe/Radar Mapper.

The first mission, the Venus Radar Mapper (VRM), is required to complete the basic characterization of the surfaces of the triad of the most "Earthlike" planets: Mars, Venus and, of course, Earth itself. Because of the scientific significance of this mission, the VRM has the highest priority. NASA hopes to

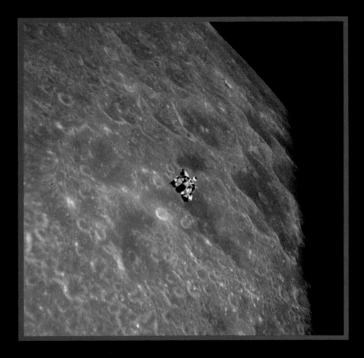

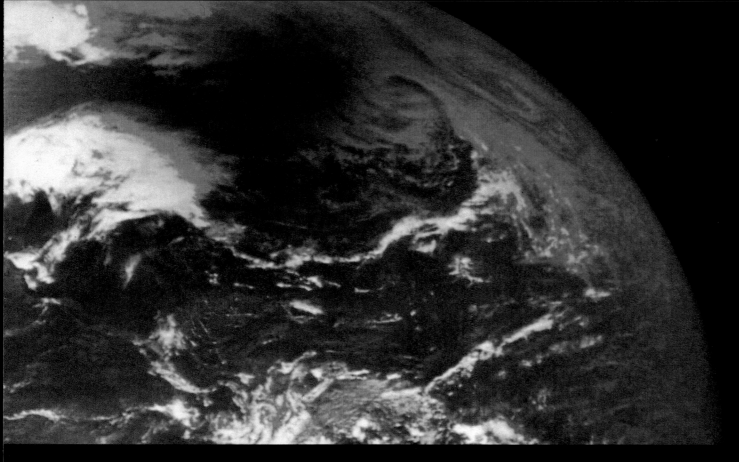

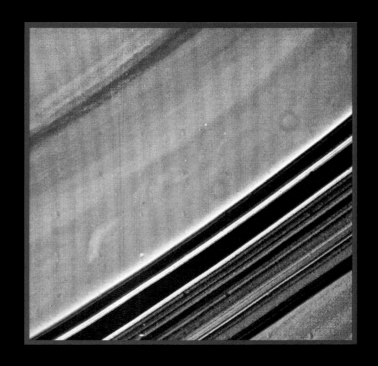

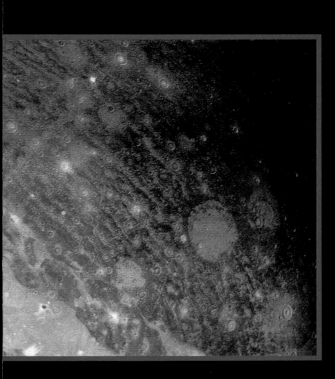

Top row, left: This photo taken from the Command and Service Module of *Apollo 16* shows a view of the Lunar Module returning from the lunar surface. *Top row, middle:* This color composite of the Jovian surface reveals the great activity taking place on Jupiter. *Top row, right:* This closeup of Saturn's rings was taken by *Voyager II* on August 20, 1981, from a distance of four million miles (6.4 million kilometers). *Bottom row, left:* This computer generated photo of Callisto, a moon of Jupiter, was taken by *Voyager I. Bottom row, middle:* This photo taken by *Voyager II* shows Ganymede, one of Jupiter's moons. The bright spots indicate recent impacts upon its surface. *Bottom row, right:* Io, another of Jupiter's moons, taken by *Voyager I*, on March 4, 1979, from less than 227,000 miles (377,000 kilometers) away.

use as many spare parts from past missions as it possibly can reduce its overall cost.

The Mars Geoscience/Climatology Orbiter is the first of a new class of Planetary Observers (PO's) recommended by the SSEC. The PO's constitute a program of "low-cost, modestly scaled, inner solar system missions" utilizing already developed, high-capacity Earth orbital spacecraft. The PO's

Opposite page: Grasping onto the Earth radiator budget satellite prior to its deployment is the Canadian-built remote manipulator system end effector. The "photo extras" in the frame are the Challenger's flight deck furnishings reflecting off the window through which the photo was taken. Left: Following the direction of controllers in the Johnson Space Center's Shuttle mock-up and integration laboratory is a human-sized robot called the EVA Retriever. The robot was paid for through the Center's discretionary fund and is an attempt to solve the possible problem of accidental separation, which might occur when crews are living and working in space for long periods of time.

should be a "level-of-effort program" similar to that of the successful early explorer series.

The third Core mission, the Comet Rendezvous/Asteroid Flyby, requires the development of the *Mariner Mark II* spacecraft—a simple modular spacecraft that the SSEC recommends be designed to accomplish, at reasonable costs, missions beyond Saturn and the inner solar system. Finally, the Titan Probe uses the existing *Galileo* craft design to study Saturn's largest natural satellite.

The recommended missions for the terrestrial planets after the Mars Geoscience/Climatology Orbiter include the following, in no particular order of priority: a Mars Aeronomy Orbiter, a Venus Atmospheric Probe, a Lunar Geoscience Orbiter and a Mars Surface Probe.

A United States Naval
Observatory photograph of
the active nucleus of a
Galaxy.

Right: The first photograph ever taken on the surface of Mars. The photo was obtained by *Viking 1* minutes after the craft landed successfully on the Martian landscape. *Below, right:* This is the first color photograph taken on the martian surface. Taken by the *Viking 1* lander, it shows that the soil consists mostly of a fine grained reddish material. However, small patches of black or blue-black soil are found deposited around many of the rocks in the foreground. *Inset:* A meteorite, believed to be of martian origin, sits in a dry nitrogen cabinet undergoing analysis in the meteorite processing lab of Johnson Space Center's planetary and earth sciences laboratory. The sample was discovered in 1981 by an Antarctic meteorite recovery team.

EETA79001, 2

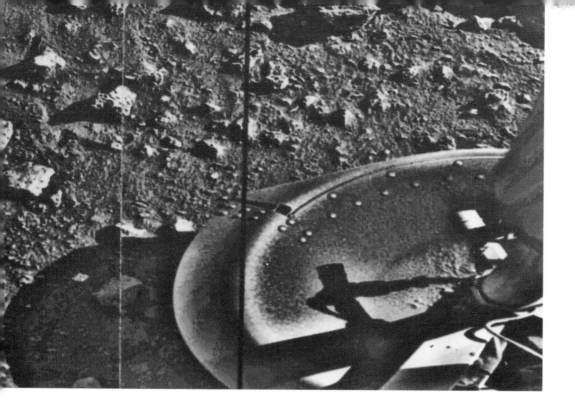

The Mars Aeronomy Orbiter is designed to explore the interactions of the planet's upper atmosphere and ionosphere using radiation and particles from the Sun, and to settle the question of whether a natural magnetic field does indeed surround Mars. The Venus Atmospheric Probe is intended to provide definitive information on the abundance of major and trace components of the Venus atmosphere. The abundance and the isotopic composition of these gases are tremendously important to scientists in their attempts to understand the origin of the Venus atmosphere. The Lunar Geoscience Orbiter will provide a global map of surface composition and other properties, as well as deciding the question of the presence of condensed water and other volatile elements trapped in the polar caps.

The Mars Surface Probe mission will establish seismic and meteorological stations and geoscience observation sites. This mission will determine the level of Mars's seismicity, analyze the surface weather data for its climatic patterns and perform geochemical and other analyses of its surface materials.

These and the other missions of the Core program are all of the highest priority in pursuing the primary scientific goals of the planetary exploration program: to reach an understanding of the present state, origin and history of the solar system, including the Earth, and the chemical history of the solar system in relation to the appearance of life. The lunar mission, like the Earth-approaching asteroid missions, also supports the program's secondary goal: the establishment of a scientific basis for future use of near-Earth resources such as water, ores and other materials that may be found on neighboring planets and asteroids.

Beyond the Titan Probe/Radar Mapper mission, the outer planets missions proposed for the Core program are a Saturn Orbiter, a Saturn Flyby/Probe and a Uranus Flyby/Probe.

Pioneer and *Voyager* observations of Jupiter and Saturn and their rings and satellites have revealed a diversity of natural phenomena highly relevant to understanding the formative and evolutionary processes of the solar system. The systematic study of the outer planets and their complex ring and satellite systems remains a major element

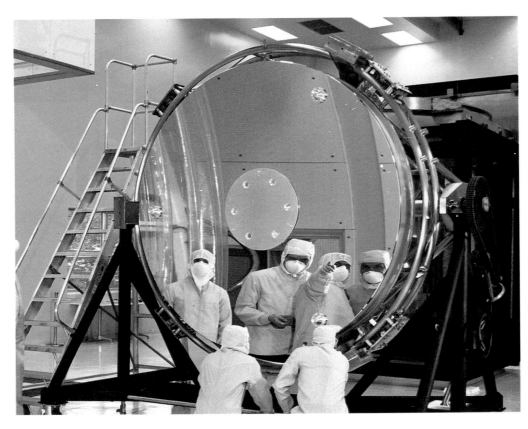

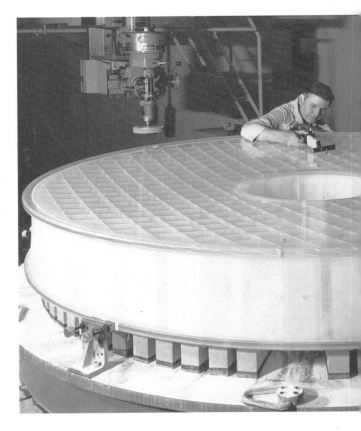

of the planetary program. The *Galileo* mission, a Jupiter orbiter and a cooperative probe project with West Germany, will address many of these goals for the planet Jupiter. After this, the next probable area of intensive study will be Saturn and its system. Characterization of the atmosphere, environment and surface of its moon, Titan, is the outer planet scientific objective of high priority. Future study will necessitate several missions, each having its own goal.

In the midst of the goals for the 1990s is NASA's current goal of developing, within a decade, a permanently manned space station. This goal, reflecting President Reagan's directive to NASA in his State of the Union Address of January 1984, is also essential to the United States space program not only to preserve a leadership role, but also to utilize the economic and scientific opportunities offered in space. The goals of the Space Station program, NASA's focus since May 1982, are to:

- Establish the means for a permanent and productive presence of people in space.

- Establish a routine, continuous and efficient use of space for science, applications, technology development and operations.

- Develop further the commercial use of space.

- Develop and further exploit the

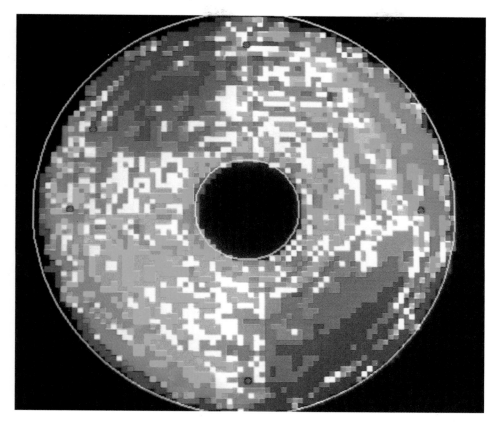

synergistic effects of the man/machine combination in space.

- Provide essential system elements and operations practices for an integrated, continuing national space capacity.

- Stimulate the mutual benefits traditionally derived from cooperation in space with allies and friends.

- Reduce the cost and complexity of living in and using space.

- Be a major contributor to United States leadership in space in the 1990s and beyond.

- Ensure that the elements of the Space Station are compatible with, and can interface with, the elements of the operational Space Transportation System (the Shuttle system).

- Motivate future scientists and technologists and provide leadership in furthering their education.

One of the major challenges of the Space Station program is to determine the synthesis of humans and machines in order to achieve the best possible human, human-tended and automated systems. It is currently unclear what role the Space Station will play in the other missions, but it is certain to augment any long-term work to be done in space.

Far left: Technicians inspect the "crown jewel" of optical astronomy, a ninety-four-inch (235-centimeter) primary mirror for NASA's Space Telescope. They wear special suits and masks to maintain the absolute cleanliness of the mirror's surface. This telescope, scheduled to be launched by the Space Shuttle, will permit scientists to peek seven times further into space than the largest ground-based telescopes can. *Middle:* A photograph of the Space Telescope primary mirror during the initial grinding phase of its fabrication by the Perkin-Elmer Corporation. The "egg crate" structure of the mirror is central in this view. *Left:* A computerized, color-coded topographic map of the surface of the Space Telescope primary mirror at the beginning of the polishing process. White represents the optimum surface shape; blue identifies the highs and lows. The pattern is based on precise interferometric measurements. *Below:* The Jet Propulsion Laboratory's Deep Space Network (DSN) which communicated with and tracks automated scientific spacecraft traveling in deep space.

WHY PLANETARY EXPLORATION?

The United States Planetary Exploration Program's purpose is to achieve a well-grounded understanding of the solar system, the planets and the Earth. There are two motivations for achieving these insights.

The first is to comprehend the solar system and its origin, one of the oldest goals known to humankind. The program's ultimate objective is to discover how basic physical laws operate to produce the world in which we live. Such comprehension would enable us to predict and control natural phenomena. Planetary science uses theory, experiment and observation to turn

knowledge of natural laws into an understanding of the world around us. A major goal of this inquiry is an understanding of the origin and prevalence of life in the cosmos. Occasionally, a human scientist can make a difference. For instance, Earth scientists had listened to astronaut accounts of subtle differences in ocean color for years. The astronauts repeatedly stressed that photographs snapped from orbit, magnificent and beautiful as they seemed, were only a pale, washed-out rendition of what could be seen by the human eye while in space. Oceanographer Paul Scully-Power, aboard a routine Shuttle mission in 1984, discovered spiral eddies (air currents) that covered the entire Mediterranean Sea, as well as cloud currents that served him as reliable tracers for tracking ocean currents. Astronomers aboard *Skylab*

An illustration of a proposed space station design. This configuration was conceived by the Rockwell International Corporation of Downey, California.

were able to use their trained eyes to create excellent drawings of the comet Kohoutek. It is also important to remember the dramatic human element in the Skylab mission in May 1973 (See Chapter 4).

The second motivation is the recognition that the solar system is the entire extended environment of humankind. What occurs on other planets in the solar system may come to pass here on Earth. There is, ultimately, no conceptual barrier to expanding the realm of human endeavors. What we are accomplishing, and will accomplish, in space will shape our next step. The Space Station, initially targeted for human habitation in the 1990s, will, without a doubt, open the gate to a wide variety of applications. While the first pioneers who have already begun to utilize the space environment for various purposes will remain mainstays in space exploration, there will emerge, in time, other brilliant men and women who will lead us toward what they see to be important in space.

Undoubtedly, the explosion of the Space Shuttle *Challenger* in 1986 sent powerful shock waves throughout the entire space community. The *Challenger* was to be the twenty-fifth flight in the Space Shuttle program and the ninth scheduled for the fiscal year 1985 (the flight was originally scheduled for July of that year). The image of the tragedy is still deeply embedded in the psyche of the nation, and its effects will most probably continue into the next decade.

At present, NASA has its hands full, rebuilding the Shuttle fleet, constructing the Space Station and maintaining a balanced program of space science and applications so that the United States and its associates can make use of the Station once it is built and in its low-Earth orbit. There is little doubt that the next decade will be difficult. Space is still a frontier and a dangerous territory similar, in some ways, to the Wild West of a century ago, or an ocean voyage before the eighteenth century. Without this far-reaching perspective, it is quite easy to forget that explorers who play a major role in shaping our knowledge of ourselves and our future often pay with their lives for their place in history. The *Challenger* accident has reminded us that we are still building our stellar highway.

Right: Large portions of the three main engines of the space shuttle *Challenger* have been recovered from the floor of the Atlantic Ocean, east of Kennedy Space Center. *Below:* Resting on the ocean floor, approximately twenty-three miles (thirty-seven kilometers) east of Kennedy Space Center, is a portion of the left-hand solid rocket booster aft segment, which contained the external tank attach ring for the destroyed *Challenger* craft. *Opposite page, top, left:* A NASA investigation photograph, as of sixty seconds into the launch. *Opposite page, top, middle:* This is a computer drawn image of a top view of the shuttle *Challenger* configuration, suggesting a possible solid rocket booster malfunction scenario, which may have been experienced immediately prior to the explosion. *Opposite page, top, right:* A NASA investigation photograph taken 76.425 seconds into the *Challenger* launch. *Opposite page, below:* In tests conducted by NASA's Johnson Space Center to study one of two proposed space shuttles escape systems, a Navy parachutist slides down a pole to exit a C-141 aircraft.

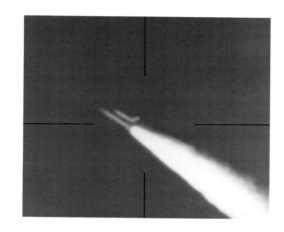
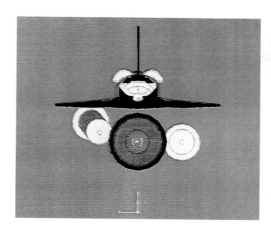
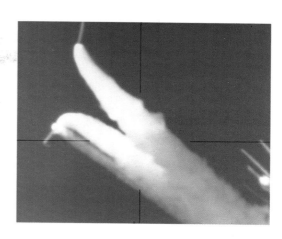

Right: A Cray-2 simulation of a particle path over a double delta wing showing vortex breakdown. *Below:* A surface particle path over a "flying wing" design. This path shows both the high and low angles of attack. *Opposite page:* A simplified grid of F-16 body surface geometry. The aircraft body is green; the wing is blue.

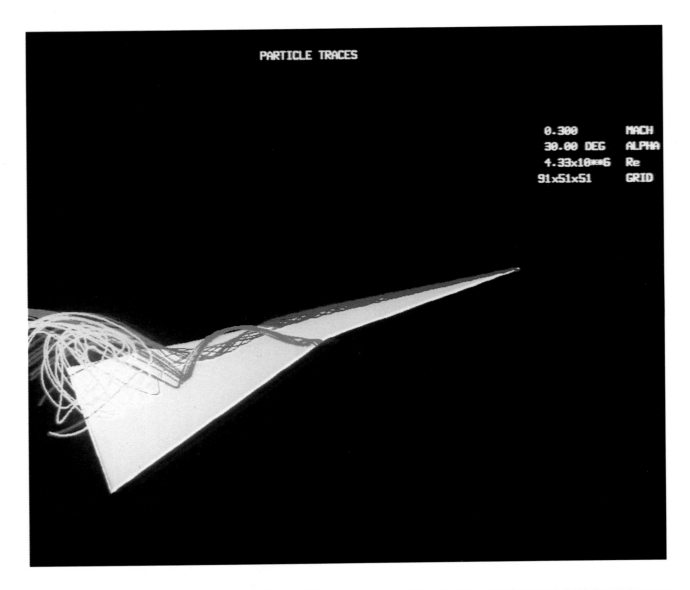

PARTICLE TRACES

0.300	MACH
30.00 DEG	ALPHA
4.33x10**6	Re
91x51x51	GRID

THE 21ST CENTURY

What will the astronauts of the twenty-first century do in space? Certainly the chores and experiments of the next century promise to be different than those of the present. If the cost of access to space can be dramatically reduced by future discoveries in technology, we can hope to see developments in key areas such as communications, astronomy, weather forecasting, Earth observations, military research, materials processing, planetary science and the search for extraterrestrial intelligence or life.

It is possible that the astronauts could be used as aggressive symbols of national pride and power; just as easily, they could be used to strive toward the goals of world peace and understanding, as in the case of the 1976 mission where an *Apollo* Lunar Module docked alongside a Soviet *Soyuz* spacecraft. This mission served as an example of détente, a litmus test of the ability of American astronauts and Soviet cosmonauts to rescue each other, a first step toward possible Soviet-American cooperation in space, and a scientific mission as well. A possible vista of the future is a joint United States/Soviet Union mission to Mars as the focus of exploration efforts by both countries. This could be a beacon of international cooperation and growth toward a new and brighter future.

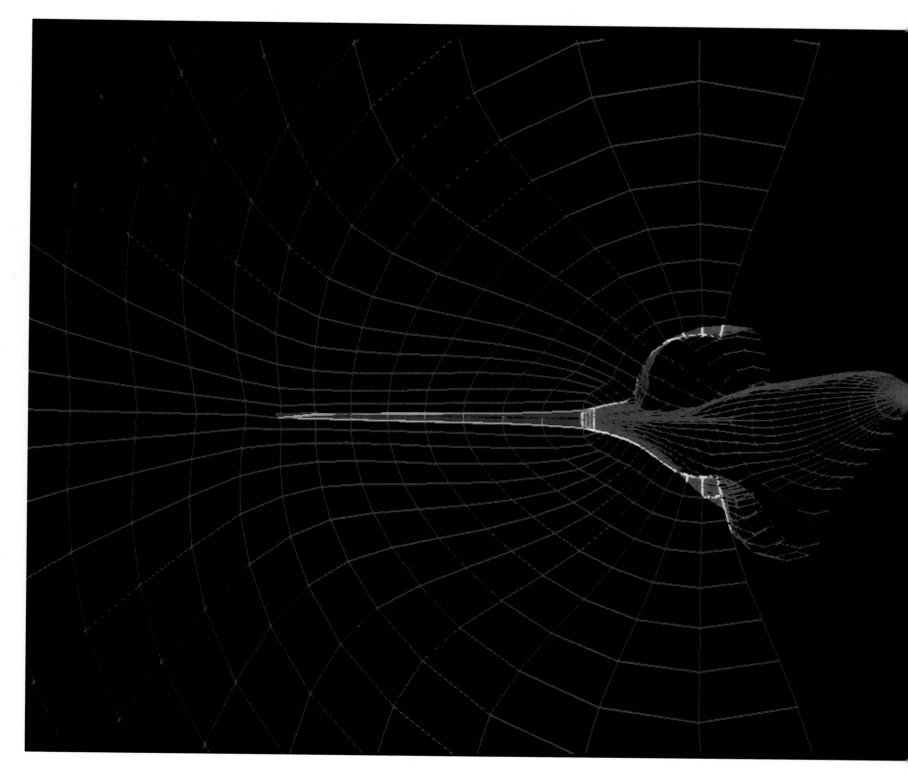

Right: A Cray-2 simulation photo of an incompressible flow over a cylinder/flat plate. The simulation is a simplification of the flow of a gaseous hydrogen around posts in the shuttle engine. *Opposite page, top:* One of a series of images from GOES-1 on the night of January 19–20, 1986, illustrating the temperature trends and patterns in an overnight freeze. *Opposite page, middle:* This image was taken by the Halley Multicolor Camera on March 13, 1986 at a distance of 12,400 miles 920,000 kilometers) from the nucleus of the comet. *Opposite page, bottom:* These two pictures of the same image of an office chair serve to illustrate a 3-D laser device instrumental in the operation of a human-size robot, used as a EVA Retriever. The left image is what the EVA Retriever would "see;" the right is comparable to what the ground controllers might see.

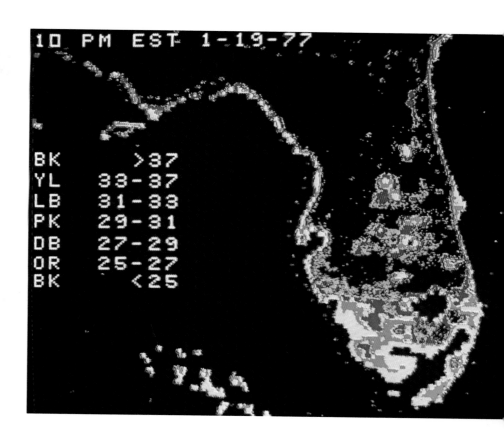

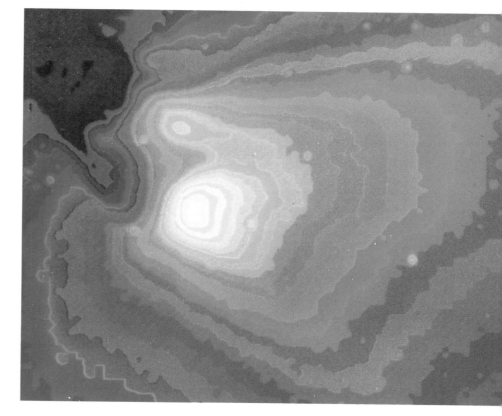

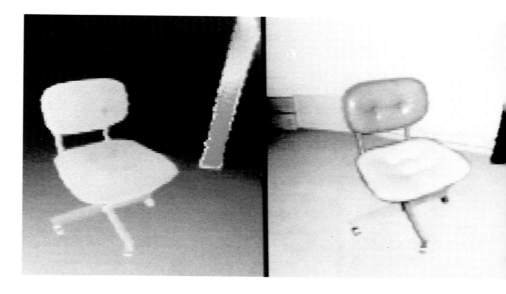

SPACE TOURISTS

Another possible role for humans in space is, of course, as tourists. An organization called Society Expeditions, based in Seattle, Washington, has already collected 265 reservations for an orbital sight-seeing tour. The company expects the tour to cost $50,000 per ticket for a twelve-hour orbital flight. The company does not yet have a launch vehicle, since the Shuttle, at $1 million per passenger, is too expensive, as well as unavailable for a decade. In years to come, however, space entrepreneurs will create bizarre trips. Anyone who has read Ray Bradbury's *The Martian Chronicles* can easily envisage droves of frantic tourists, armed with cameras and straw hats, disembarking from a spacecraft lined with naugahyde seats, bright orange decor and large velvet photos of men and women who valiantly served the space program in the past. As these "Martians" trek toward hotels with names like The Desert Sun and Earthview Arms, it will seem like eons ago that men were locked into small, tight capsules, shot into space with no creature comforts, and then, after splashdown, waited for outside personnel to unbolt that hatch before they

A view of the Eastward, showing the east half of the island of Java in Indonesia. The photo shows the volcanic origin of the island chain. To the east of Java are the volcanic islands of Madura, Bali, Lombok, Sumbawa and Sumba.

could climb out. Indeed, such things as space honeymoons, motels, low-gravity swimming pools and saunas, orbiting gambling rooms and lunar golf courses may become commonplace.

Will we ever be able to travel beyond our solar system? The past thirty years have taught us to use the word "impossible" with the utmost caution. Human travel beyond our solar system is, nevertheless, a sobering concept. Even the most forward-thinking of today's optimists do not expect humankind to make such a voyage within our generation.

Yet, the Chinese have an old adage that says a thousand-mile journey begins with a single step. In the beginning, the Chinese took that first step forward, as did the Wright brothers; men such as von Braun and Goddard dreamed of launching a rocket to the Moon—such a fanciful concept as that, two generations ago, certainly looked as bleak as the idea of interstellar travel appears today. The human mind, naturally restless, is not easily discouraged by obstacles. In fact, the steeper the obstacles, the greater the challenge for humankind to climb over them. With imagination, we can view a not-to-distant future where the frontier of space is filled with vast opportunities. NASA has helped to point not only the United States but the entire world toward that new frontier.

*W*e shall not ascribe the origins of the
            universe to inanimate bodies...
            nor make one Reason
            and one Providence...
*T*he harmony of the world is...
            like that of a bow or a harp,
            alternately tightened and relaxed.
*T*he planets are watchful gods...

                                —Plutarch

SOURCES FOR FURTHER READING

Baker, David. *The Rocket: The History and Development of Rocket and Missile Technology,* New York: Crown Publishers, 1978.

Baker, Wendy. *Nasa: America in Space,* New York: Crown Publishers, 1986.

Colby, C.B. *Beyond the Moon: Future Explorations in Interplanetary Space,* New York: Coward, McCann & Geohegan, 1971.

Collins, Michael. *Carrying the Fire,* New York: Farrar, Straus and Giroux, 1974.

Cosmic Search. Delaware, Ohio: Cosmic-Quest, Inc., 1986.

County Libraries Group. *Reader's Guide to Books on Space Technology and Exploration,* Newton, England: The Group, 1971.

Emme, Eugene M., ed. *History of Rocket Technology,* Detroit: Wayne State Press, 1964.

Klass, Phillip J. *Secret Sentries in Space,* New York: Random House, 1971.

Lay, Beirne. *Earth Bound Astronauts,* Englewood Cliffs, NJ: Prentice-Hall, 1971.

Lewis, John S. and Ruth A. Lewis. *Space Resources: Breaking the Bonds of Earth,* New York: Columbia University Press, 1987.

Ley, Willy. *Rockets, Missiles and Men in Space,* New York: Viking Press, 1968.

Newell, Edward. *Beyond the Atmosphere: Early Years of Space Science,* Washington, D.C.: Scientific and Technical Information Branch, NASA, U.S. Government Printing Office, 1983.

Pioneering the Space Frontier: The Report of the National Commission on Space. New York: Bantam Books, 1986.

Planetary Exploration Through the Year 2000: A Core Program. Washington, D.C.: NASA, U.S. Government Printing Office, 1983.

Von Braun, Wernher. *First Men to the Moon,* New York: Holt, Rinehart & Winston, 1958.

Von Braun, Wernher and Frederick Ordway III. *The Rocket's Red Glare,* Garden City, NY: Anchor Press, 1976.

Weber, Ronald. *Seeing Earth: Literary Responses to Space Exploration,* Athens, OH: Ohio University Press, c. 1985.